VOSKUM LIBR

MW00474601

Living with Fire

Living with Fire

*Fire Ecology and Policy for the
Twenty-first Century*

Sara E. Jensen
Guy R. McPherson

UNIVERSITY OF CALIFORNIA PRESS

Berkeley   Los Angeles   London

4/09

University of California Press, one of the most distinguished
university presses in the United States, enriches lives around
the world by advancing scholarship in the humanities, social
sciences, and natural sciences. Its activities are supported by
the UC Press Foundation and by philanthropic contributions from individuals and institutions. For more information, visit www.ucpress.edu.

University of California Press
Berkeley and Los Angeles, California

University of California Press, Ltd.
London, England

© 2008 by The Regents of the University of California

Library of Congress Cataloging-in-Publication Data

Jensen, Sara E., 1978–
   Living with fire : fire ecology and policy for the twenty-
first century / Sara E. Jensen, Guy R. McPherson.
        p.   cm.
   Includes bibliographical references and index.
   ISBN 978-0-520-25589-0 (cloth : alk. paper)
      1. Wildfires.   2. Fire ecology.   3. Fire management—
Government policy—United States.   I. McPherson, Guy R.
(Guy Randall), 1960–   II. Title.

   SD421.J46   2008
   363.37 9—dc22   2007044589

Manufactured in the United States of America

17   16   15   14   13   12   11   10   09   08
10   9   8   7   6   5   4   3   2   1

This book is printed on Cascades Enviro 100, a 100% post-
consumer waste, recycled, de-inked fiber. FSC recycled certified and processed chlorine free. It is acid free, Ecologo
certified, and manufactured by BioGas energy.

CONTENTS

ILLUSTRATIONS

ACKNOWLEDGMENTS

Guy McPherson appreciatively acknowledges financial support from the USDA Forest Service Rocky Mountain Research Station and the use of the Nature Conservancy's Lichty Ecological Research Center while researching and writing this book. We also thank Courtney Sherwood, Richard Halsey, and three anonymous reviewers for their thoughtful and insightful comments. Thanks above all to Chris and Sheila for their patience, generosity, and support.

Introduction

Uncertainty and Change

We live in a time characterized by rapid, comprehensive, and often overwhelming change. Social, cultural, legal, and physical landscapes are changing. Ecosystems, economies, and even the climate are shifting in unimaginably vast and complex ways. In such a world we, the authors, have noticed a marked tendency among human societies to deal counterproductively with these changes, to fail to cope with the uncertainty and variability that are inherent to our modern lives. This cannot be surprising, but we hope it is not inevitable.

Near the end of the twentieth century, the futurist Alvin Toffler described some of the causes of our societal impotence: "In describing today's accelerating changes, the media fire blips of unrelated information at us. Experts bury us under mountains of narrowly specialized monographs. Popular forecasters present lists of unrelated trends, without any model to show us their interconnections or the forces likely to reverse them. As a result, change itself comes to be seen as anarchic, even lunatic."[1] The result of this chaos has too often been that humans fail to distinguish between the factors that we can and cannot affect in our lives. Faced with a confused flurry of insurmountable problems, we rarely manage to focus our attention on taking even the most basic steps toward mitigation. Instead of attending to the fundamental, approachable problems that we have the ability and knowledge to solve, we strive in vain to reverse inevitable natural forces. We respond to our fear of the normal cycles of flood and drought by building colossal waterworks that allow people to live and thrive in floodplains, deserts, and coastal areas, but we ignore the small-scale, long-term planning needed to protect those people when hundred-year droughts and floods overwhelm the system. We focus our medical system on avoiding death at any cost, even while many individuals in our nation live in ill health, falling prey to easily preventable diseases. We attack problems with billions of dollars, high-tech "fixes," and a determination to eliminate all risk, and in the process we undermine more meaningful, realistic, and affordable approaches to solving them.

Thus, in our confused attempts to solve each problem completely and eliminate all risk, we defeat our own interests. Our determination to solve a problem permanently and completely—which is often an unrealistic goal—can blind us to effective mitigation. Ultimately, we must admit that the forces and changes that plague us cannot be completely prevented or solved. Paradoxically, this very realization can help us find and apply realistic ways of coping with and mitigating these problems.

In the western United States today, as in many other regions, wildland fire presents just such an opportunity.[2] Fire has been a part of nearly all the world's ecosystems for millennia. It plays a crucial and irreplaceable role in the ecosystems that support all life, but it understandably provokes fear in humans. Fire is both inevitable and ubiquitous. To put an end to wildland fire in the West would be disastrous; indeed, even the limited scope on which we have controlled it so far has led to devastating ecological consequences. But as necessary as fire is to maintaining healthy ecosystems, it can also threaten human lives and livelihoods in a way that any society would find unacceptable. Therefore, it is a problem that must be mitigated rather than solved. We cannot extinguish fire permanently, so we must find a way to cope with it.

Interestingly, we have arrived at the wildland fire "crisis" precisely as a result of the media deluge Toffler describes. Humans have always struggled to live with fire and its inherent risks, but in the late twentieth and early twenty-first centuries, the media and their "blips of unrelated information" have increasingly influenced our understanding of these risks—and not always accurately. This began with the fires at Yellowstone National Park in 1988. For the first time, a major fire was also a major national media event. The fires, which burned across more than 1.5 million acres, provided gripping images of massive firestorms and charred forests. The nation was transfixed by the story. Ultimately, though, this fascination probably had more to do with several key developments in media technology than with the Yellowstone Fires themselves. From an ecological perspective, the 1988 fires were not particularly surprising or unique, nor even unusually destructive. But the media had changed. The satellite news truck, which allowed local television stations to broadcast from any place with vehicular access, was developed in 1984. In 1985, RCA launched a new type of communication satellite that made it possible for local stations to broadcast using relatively small, inexpensive equipment. RCA even gave this equipment to many local stations to help make sure its satellite would be used. Local stations soon recognized that broadcasting news on location increased their prestige and their viewership. Reporters could now travel far afield with fly-away packs of equipment that fit into a few suitcases and could be assembled for broadcasting within minutes, all at the relatively low cost of less than twenty dollars per minute.[3]

Since 1988, large wildfires nearly always have appeared on the local or even national evening news, regardless of their relevance to the lives of the audience. Just as viewers are bombarded with true crime stories and other horrific (but rare) events, the nightly news during the summer fire seasons reliably presents a bleak view of wildfires destroying the beloved American West. One fire after another is given its thirty seconds of nightly national attention, and the audience receives, not information, but images of destruction and an outline of the causes and results of the fire in fewer than fifty words.

As Toffler's work suggests, the general public has been unable to build a cohesive, coherent understanding of fire based on this chaotic mosaic of images and sound bites. Scientists, meanwhile, focus on the role of fire in a particular system at a particular time but rarely communicate their results in a meaningful context to the public or land managers. Occasionally, a catchphrase appears that seems to show the big picture, but experts who disagree with it will quickly contradict it. In today's sound-bite-driven world, these stories must be told as quickly as possible, and in seeking balance, journalists often present extreme perspectives briefly rather than a comprehensive, nuanced assessment.

This is how today's fire "crisis" has been built: Media attention, misinformation, and miscommunication have transformed a natural, inevitable, and highly beneficial phenomenon into a hopeless quagmire. The public demands that the problem be solved. Billions of dollars are poured into making forests and people "fire-safe," but the fires seem to keep getting worse. Every summer we watch another raging inferno, another community in flames. Viewers are understandably shocked. What can be done? How can such meaningless destruction be stopped?

The answer is, it cannot be. Wildland fire is not going away. It is time we learn to live with it. Fire is not a war, and an absolute victory is impossible. But to accept reality is not to accept defeat. For perhaps the first time in this nation's history, we know the basics of how to live with fire. We know a great deal about how fire works, and we know how to mitigate its effects in a way that can improve our lives in myriad ways. We can use fire to maintain and increase biodiversity, to protect our water supply and other natural resources, and to meet the many stewardship goals that we set as a society and as landowners. At the same time, we know how to make our homes as safe as possible from fire and to prepare for fire where it is inevitable. Our formal knowledge about fire, admittedly incomplete, is still vast and growing.

Unfortunately, much of that knowledge is difficult for the average American to access. It is scattered among expert sources, each of them viewing the fire problem from a quite narrow perspective. Fire ecologists in the academic world necessarily limit their conclusions to one or a few ecosystems or processes. They are specialists, and fire ecology is an almost unimaginably broad field of study. Many employees of federal land management agencies have a relatively broad understanding of fire. But both academic and land management expertise are generally derived and expressed in ways that are opaque to the nonexpert. They can be found in jargon-ridden (not to mention expensive) academic journals, inaccessible or obscure governmental reports, and the even more obscure literature of both realms—conference proceedings and minor agency-specific reports. Very little meaningful information ever filters beyond the realm of wildfire experts to the general public, or to academics and experts in other, even closely related, fields.

In this book, we argue that the current state of knowledge is more than sufficient to set us on the path to living peacefully and productively with fire. In order to reach this path, however, we must corral and organize our scattered knowledge. We must step back from misinformation and overcome miscommunication. When we do this, we can begin to understand where we are, how we got here, and where we need to go.

Our aim in writing this book is to make critical knowledge on wildland fire available to nonexperts. These include, among others, residents of fire-prone areas, policy makers, and fire managers. These groups of people, who lack easy access to knowledge about fire are, arguably, the very groups most affected by it. We also want to provide an interdisciplinary perspective for fire experts by supplying scientists and managers with much-needed policy information, and policy experts with equally important scientific knowledge. We show these audiences how an improved understanding of fire's role in ecosystems can help us to end our battle against it, and to learn to live with it and use it to our own ends. We also provide some specific recommendations for future approaches to fire management and policy.

The first chapter constitutes a state-of-the-knowledge report: we summarize the current scientific understanding of fire. Wildland fire is incredibly complex and unpredictable, and it is influenced by countless interrelated and synergistic factors. Meanwhile, our knowledge of it is vast but, as we have said, terribly disorganized and inaccessible. The remainder of chapter 1 outlines the details of a new paradigm of wildland fire that recognizes the important and complex role that fire plays in western ecosystems. Chapter 2 challenges earlier paradigms by examining several of the less-recognized factors that we now know greatly influence wildland fire. In chapters 3 and 4, we discuss two misguided attempts to solve the fire problem, one historical and one current. Current popular views of fire, like those that preceded them, have called for attempts to eliminate fire from natural landscapes. We show how these attempts have exacerbated the very problems they aimed to solve, even while undermining the ecological and political systems that support us. Finally, chapters 5 and 6 recommend new directions for fire management, including policy design and implementation, based on a more appropriate interpretation of economic and ecological realities.

Most North American ecosystems have burned periodically since the glaciers retreated more than ten thousand years ago. Fire was once the dominant ecological process in western forests, but since European settlement its occurrence and spread have been vastly reduced. Toward the end of the twentieth century, however, fire began to make a comeback. Perhaps nature is serving notice: our days of controlling fire are over. For some, this may seem like an apocalyptic prediction. However, as authors, researchers, and citizens, we are optimistic that a strategy for dealing with the inherent complexity of wildland fire can help the

United States rebuild critical ecological and political processes and, eventually, the nation can learn to live with fire.

As we write this in early 2007, we acknowledge that we might well be accused of fiddling while Rome burns. In the face of massive challenges that confront our country, seemingly on every domestic and foreign issue, it has been difficult to focus on such a narrow, even apparently arcane, topic. We suggest, though, that the solutions we present here have the potential for much broader application. Our general strategy holds true in many dilemmas. A problem seems insurmountable when we, as a nation, are unable to see the whole of it. In light of our incomplete knowledge, the perfect solution we seek is unattainable; meanwhile, we hold in our hands the very tools needed to mitigate the problem and reach a compromise solution. Perhaps as we learn to live with fire, we can learn to seek moderate solutions in other realms as well. We certainly hope so.

ONE

Wildland Fire in the West

The Big Picture

Scientific understanding of wildland fire has grown exponentially during the past few decades. Fire ecology has developed into a viable field of study. Scientists and land managers increasingly recognize the complexity and uncertainty involved in fire management, and alternatives to fire suppression have gained popularity. Today we can identify three major methods for managing fire: First, we can suppress it directly, to the best of our ability. This is the method that has predominated since the mid-twentieth century. We also can manage it indirectly, by physically removing the fuels necessary for fires to ignite and spread. Physical fuel reduction is most often accomplished by selective cutting, although the use of cattle or goats for removing finer fuels also has many proponents. Finally, we can use fire to our advantage, either by setting prescribed fires or by letting lightning-sparked fires burn.[1] Fire use also changes the behavior of future fires by removing fuels, and many land managers and policy makers consider fuel reduction through physical removal to be roughly functionally equivalent to reduction by fire. As we will see, however, the results of these two methods can be very different in practice. In any case, fire suppression, fuel reduction, and fire use are the primary tools available to fire managers today. We discuss these tools and their applications in much greater detail in chapter 5.

Throughout most of the last century, fire use has been highly contentious. The idea of using fire to accomplish specific management goals is quite old (many Native American cultures were frequent burners, as were early white settlers), but it has only recently gained ground as an alternative to suppression, at least in the eyes of the federal government. In what became known as the Leopold Report, several ecologists in 1963 famously endorsed the idea of letting fires burn under well-defined conditions.[2] The National Park Service accepted the practice as a management technique shortly afterward, but was slow to put it into widespread use.

From publication of the Leopold Report through the 1980s, debate raged over the wisdom of allowing some lightning-ignited fires to burn.[3] Research on historical fire cycles and fuels suggested that it might be beneficial in some areas to allow fires to burn unchecked in circumstances where they were unlikely to grow large and intense. Land managers began tentatively allowing fires to burn, especially early and late in the fire season and during relatively cool, humid, and calm conditions. They hoped that allowing numerous small fires to burn might head off very large, intense fires by keeping fuels from building up. As the use of those managed fires gained in popularity, this practice also promised land managers an opportunity to safely burn up more fuels every year without having to budget for costly prescribed fires.

By the late 1980s, the National Park Service seemed to be doing everything right at Yellowstone National Park. In response to scientists' recommendations, particularly the influential Leopold Report, Yellowstone had developed a policy that allowed some lightning-caused fires to burn unchecked. Managers recognized the key role fire played in park ecosystems, especially in forests dominated by lodgepole pine (*Pinus contorta* var. *latifolia* Engelm.), a species that both facilitates and depends upon fire. (Some populations of lodgepole pine produce closed cones that are not viable until they are opened by fire.)[4] The Park Service had allowed 235 fires to burn 33,759 acres in Yellowstone between 1972 and 1987; all of them were extinguished naturally. Only a few ever covered more than a hundred acres.[5] It looked for some time as if the reintroduction of fire might be a fairly straightforward process there, and using naturally ignited fire as a management tool started to gain political acceptability, both within Yellowstone and nationally.[6] Managers believed that fires were unlikely to grow large except in Yellowstone's old-growth pine and older spruce-fir forests, and these occurred in wilderness areas that seemed large enough to contain nearly any fire. In retrospect, it seems likely that a string of wetter-than-average summers through the early and mid-1980s unduly influenced these perceptions.[7]

As the summer of 1988 began, the Park Service's assumptions were put to the test. After an unusually wet spring, the normal summer rains virtually stopped. The Yellowstone area entered a severe drought and grew increasingly parched as dry thunderstorms began to spark fires. Early in the summer, lightning-strike fires were allowed to burn, provided they fit within preestablished parameters. By July 15, drought conditions had worsened enough that the Park Service began to aggressively suppress all new fires. After July 21, the service actively fought all existing fires.[8] A total of 248 fires started that summer in the greater Yellowstone area, and several of them burned through the summer and into the fall. High winds and dry fuels made firefighting a losing battle, and the situation seemed to many observers to be totally out of control. The fires became manageable only when snow started falling in mid-September. By that time, fires had burned across about 1.5 million acres of the greater Yellowstone area, including about 36 percent of the park itself. The National Park Service reported that "more than 25,000 firefighters, as many as 9000 at one time, attacked [the fires] at a total cost of about $120 million."[9]

Media coverage of the fires was emotional, hyperbolic, and unrelenting. It also was very often misleading. Live coverage of flames and charred forests was everywhere. The alarming images and rhetoric, along with outright disinformation, led the public to believe that intense fires had completely destroyed the park's beauty and wonder. They also suggested that the Park Service had, based on a contentious and misguided theory of forest management, allowed America's most prized national park to burn to the ground. The public had come to expect quick and effective suppression of all fires, and there was little patience for the argument that some fires simply couldn't be suppressed. Even as park officials were spending three million dollars per day on fire suppression, they continued to insist to the media that only winter could extinguish the fires.[10] Still, most Americans seemed determined to believe that modern technology, coupled with sufficient personnel, should be able to extinguish any forest fire.

In the wake of the fires, many people, including some scientists, land managers, and members of the public, believed that the soil had been sterilized and the forest effectively destroyed for decades to come.[11] The Park Service's let-burn policy became a major target of public outrage, and interest in "natural" fires waned. Some fire scientists argued that a program of careful prescribed burning would have prevented the whole catastrophe, while many in the media argued for a return to total suppression.[12]

It wasn't long, though, before it became clear that Yellowstone had not simply been converted into a wasteland and a public testament to Park Service incompetence. The following spring, wildflowers bloomed on the blackened ground. Within ten years, even the most severely burned forests were scattered with small lodgepole pine seedlings.[13] Today we know that the burned landscapes have largely been recolonized by native species, that the fires led not to a homogeneous landscape but to a diverse mosaic of burned and unburned areas, and that the snags and downed logs created by the fires have decayed and formed a rich source of organic matter that boosted productivity.[14] We also know that many of the management strategies previously suggested for the Yellowstone lodgepole pine ecosystem—from total suppression to reducing fuels and creating a regime of frequent, minor ground fires—are misguided, oversimplified, and ultimately futile.

The 1988 Yellowstone Fires have many lessons to offer. First and foremost, they show us that our society has a pervasive and unfortunate tendency to misjudge wildland fire. We underestimate fire's role in ecosystems, and we seriously overestimate humans' ability to control it. This is a lesson that nearly any wildland fire, regardless of size and severity, can teach us. In later chapters, we examine our government's ill-fated attempts to gain control over fire and show the inevitable failures of that path, and we offer suggestions for living productively with fire, both as a society and as individuals. First, however, we examine in some detail the beautiful, astonishing complexity of wildland fire and the inherent uncertainty involved in living with it.

MAKING SENSE OF COMPLEXITY

Wildland fire is an unimaginably complex and unpredictable phenomenon influenced by countless interrelated factors, and as such, it does not submit easily to human control and planning. During the Yellowstone Fires, for example, a totally unpredictable change in winds caused a single fire, brought partially under control by firefighters, to explode, growing from 68,000 acres to almost 250,000 acres in just sixteen hours. A few days later, rain and snow brought the fire back under control.[15] Fire in its wild state is notoriously difficult to anticipate, much less control.

At the most basic level, however, we understand fire quite well. Fire requires only three things: heat, oxygen, and fuel. More specifically, it requires a heat source for ignition, an adequate supply of oxygen, and dry, fine fuel sufficient to ignite and then carry fire.[16] The environment around a wildfire transforms these three requirements into a practically infinite set of variables. Ignition can be lightning or human caused. Oxygen can be supplied under a variety of wind patterns. Winds may be calm or gale force, stable or rapidly shifting, and caused by massive, slow-moving fronts or sudden, unexpected changes in the jet stream. Large fires create their own winds and weather systems. Fuels can be fine or heavy, green or dead, and can become extremely desiccated during drought conditions. At one point early in the summer of 1988, moisture levels of some of Yellowstone's fine and dead fuels were lower than that of kiln-dried lumber.[17] Some plants, including several dominant species in chaparral communities, contain highly flammable volatile oils. Others, including many nonnative grasses now dominant in the southwestern United States, have a life cycle and physical structure that predispose the species to carrying fire. Moreover, the flammability of a fuel type can be radically altered by factors such as slope and exposure. The result is an amazingly diverse universe of unique combinations of ignition type, fire weather, and fuel type. Any single variable is capable of completely altering a single fire's behavior or even a specific ecosystem's entire fire cycle.

The term *fire regime* is used to express the idea that any given ecosystem has evolved with fires of a certain kind—a certain frequency, seasonality, intensity, and extent. Fire ecologists often aim to describe (and fire managers aim to re-create) the "natural" fire regime for a specific ecosystem. The term is useful to the point of being indispensable, because it gives us a framework for understanding the patterns common to fires in many different systems. But the idea of a clearly defined fire regime can also oversimplify the complexity of fire and its variations across space and time, even within a single ecosystem. The term obscures the diversity and unpredictability of fire as it interacts with and changes diverse landscapes. Even while we describe the fire regime for ponderosa pine forest or desert ecosystems, therefore, it is important to keep in mind that any such description is an oversimplification. In reality, fire regimes change over time (and may change even more quickly under human influence), just as they vary among regions, among different mountain ranges, and even from slope to slope or forest to forest.

Fire scientists describe both individual fires and fire regimes according to several different traits, including frequency, intensity, seasonality, extent, and weather. As a result, fire regimes can be defined and classified in a surprising variety of ways. In order to plan for and manage fire appropriately, it is important to be clear in our definitions. When scientists and land managers have classified fire regimes, they have typically focused almost exclusively on the frequency of fires in a given ecosystem. An ecosystem's fire regime is usually described in terms of fire-return interval, the estimated average period between fires. Defining fire regimes by their return interval alone can be problematic, because this reduces a dynamic trait to a single range of numbers. For example, a semiarid "desert" grassland of the southwestern United States might be described as having a fire return interval of seven to ten years, which would suggest to land managers that fires "naturally" occur at this interval. In fact, managers often see fire-return interval as a goal to be met and plan prescribed burns or fuel treatment projects in response to these figures.

The fire-return interval can be a misleading figure, however. Historical fire frequency is difficult and in some ecosystems virtually impossible to determine. Pinpointing it requires a great deal of site-specific information, which is most commonly gathered through dendrochronological (tree ring) analyses. When a given tree survives a fire, its rings show scars. In forests with many old trees that have survived multiple fires, scientists can reconstruct a hypothetical fire history based on these fire scars. However, the accuracy of this method is somewhat uncertain. A fire may burn through an area and not leave scars on all trees. Fires often burn in a spotty "mosaic" pattern, and even a tree that is charred may not necessarily scar.[18] In any case, many forests, and certainly most grass- and shrublands, lack the large, old trees needed for this kind of analysis. Notes from General Land Office

Surveys or historical photographs are sometimes used to help build a historical record. On the whole, however, fire-return intervals should be viewed with a healthy dose of skepticism.

Fire-return intervals also change over time in response to climate and land-use patterns. For example, the southwestern desert grasslands themselves may be an artifact, at least in part, of intentional or unintentional burning by Native Americans, in which case "natural regime" is a misnomer. A more accurate description of fire frequency might be one stating that fires occurring every seven to ten years will tend to maintain a grassland ecosystem, whereas areas characterized by less frequent fires will likely become dominated by woody plants.[19] Some systems also have mixed fire regimes in which frequent minor fires are interspersed with infrequent extreme fires. Classifying fire regimes according to frequency effectively obscures these important variations. Fire-return intervals can be misleading also because they make it easy to ignore other regime traits that may have even greater impacts on both humans and ecosystems.

Fires and fire regimes are also often described in terms of intensity. At its most basic level, *fire intensity* describes the heat released by a fire. The intensity of any given fire varies over both time and space.[20] *Intensity* also tends to be used as a shorthand descriptor for other traits closely associated with high- or low-intensity fires. For example, intense fires are often also very large fires, partly because hotter fires are more difficult to control. Similarly, fires may be described as intense because they display behaviors associated with large, hot fires, such as spotting, fire whirls, crowning, and long, fast runs.[21]

Ecosystems are further characterized by a climate conducive to fire during particular seasons. For example, the southwestern

United States experiences a hot, dry period nearly every year between early April and late June, and this spring drought is broken by "monsoonal" thunderstorms. Thunderstorms usher in the summer when fuels are very dry; lightning from these storms historically ignited fires, and it still accounts for many wildfires. Storms that contain little precipitation are particularly likely to spark fires. As might be expected, native species in the region are adapted to fires that occur during this period. Land managers, however, often prefer to light prescribed fires during early spring or late autumn to reduce the likelihood that a fire will rage out of control. Native species are poorly adapted to fires that occur "out of season" relative to their evolutionary history, and such fires expose these organisms to conditions as unusual—and potentially as deadly—as a midsummer snowstorm. The season in which a fire occurs can affect plants' ability to produce seeds and take advantage of seasonal precipitation for growth and reproduction.[22] It can also dramatically influence fire behavior. Winter fires may be so low in intensity that they hardly reduce fuels in ecosystems where summer fires have historically predominated.[23] Therefore, fire use that does not take into account an ecosystem's historical fire season can result in a shift in dominant species, an increase in nonnative species, and even the extirpation of rare or particularly vulnerable organisms.[24]

Fire extent is another important trait to consider when measuring and describing fires: large and small fires tend to have very different effects on ecosystems. Large fires, especially intense ones, leave behind large "islands" of habitat that are structurally different from the surrounding area. Small or immobile species may be very slow to recolonize large burned patches, a circumstance often described as a major fault of large, intense fires, because the area is seen as "dead." But it is not dead at all. Birds, butterflies, and large ungulates may begin using even very large, intensely burned areas within a few hours after a fire has passed. Smaller and less intense fires create a mosaic of effects and may contribute greatly to the diversity of landscapes and thus to biological diversity.[25]

Of all the factors that influence wildland fire, weather is among the most profound and complex. Weather and fuels interact in important but often unpredictable ways, which helps explain why humans find it so difficult to influence wildfire behavior. Fire managers have long recognized the importance of weather variables in planning and conducting prescribed fires and in suppressing fire. Plans for prescribed fires include a strict range of suitable weather conditions outside of which the fire cannot be ignited. But weather doesn't follow human plans; a recent government study reported that weather concerns accounted for 40 percent of the delays on fuel-reduction projects at sites visited by researchers.[26] A basic understanding of how weather affects different kinds of fire regimes is crucial to the effective management of fire-prone and fire-dependent ecosystems.

Weather largely determines fuel moisture and thus the volatility of fuels. Hot, dry, windy weather over an extended period drives moisture from wildland fuels just as it would from an open pan of water. Small-diameter fuels lose moisture much more rapidly than larger fuels, because grasses and smaller leaves and branches have more exposed surface area from which moisture evaporates. As a result, grasslands and shrublands become tinderbox dry after only a few hours of desiccating winds. It takes considerably more energy and time to dehydrate large trees in shady forests.

A close relationship exists also between a given ecosystem's response to weather and its fuel structure, in terms of the relative level of canopy cover. *Canopy cover* describes the amount and distribution of leaf area in an ecosystem and can range from very open—or even nonexistent, such as in a few of the world's deserts—to completely closed, such as in dense forests where there is no space between tree crowns. In grasslands and the least-dense shrublands and forests, fire is usually carried at ground level and through herbaceous fuels. As a result, fires tend to be less intense, although they may still spread very quickly. Where the forest or shrub canopy is more closed, undergrowth is less grassy and tends to consist of young shade-tolerant trees and shrubs of various sizes.[27] These "ladder" fuels often carry fire into the canopy, where it burns in very hot, intense, and rapidly spreading crown fires.[28]

Although all fires are driven by the same general forces of fuel, ignition, topography, and weather, the relative importance of these factors differs somewhat among ecosystems. In recent years, a useful new paradigm has emerged for describing fire regimes. While it cannot encompass the full variability of fire, this paradigm represents an important effort to move beyond the assumption that regimes are defined by the average frequency of fires. Today many fire ecologists describe three broad classes of fire regimes: low, mixed, and high severity.[29] Low- and high-severity fire regimes represent opposite ends of a continuum. Each designation corresponds to several ecosystem types and is generally defined by evidence of an ecosystem's fire history before suppression was widespread. In this case, the term *severity* indicates a number of interrelated traits, including fire intensity and behavior, fuel type, and the influence of weather. We will look at each of these regimes in turn.

The dynamics of low-severity fire regimes are relatively well known because of exhaustive studies of low-severity-regime ponderosa pine forests in the southwestern United States. The Ecological Restoration Institute at Northern Arizona University has been especially active in unraveling the complex fire regimes of these systems. Low-severity fire regimes are characterized by open canopy structures, which produce and are maintained by relatively frequent, low-intensity fires. Low-severity regimes generally exist in ecosystems that experience annual seasonal droughts, where weather and fuel moisture conditions are favorable for fire nearly every year. Fires occur when fuel loads build to a critical threshold. Because fires in low-severity regimes tend to be driven more by the accumulation of fuels than by dry weather, it is possible to manage fire in them somewhat by manipulating fuels.[30]

This is not to say that low-severity regimes are immune to weather and climate influences. In low-severity-regime forests of the southern Rockies, for example, synchronous large fires tend to occur when fluctuations in the El Niño–Southern Oscillation (ENSO) produce a wetter-than-average winter and spring one year (known as an "El Niño year") and drier-than-average conditions the next (a "La Niña year").[31] Large amounts of vegetation accumulate during the wet year and are quickly desiccated during the dry summer that follows. As a result, La Niña years are often associated with busy fire seasons across the southern Rockies.[32]

High-severity fire regimes, conversely, are often driven by short-term weather patterns more than by the accumulation of fuels. Fires can be frequent or infrequent (even centuries apart, as with the lodgepole pine forests of Yellowstone National Forest),[33] but they tend to be intense and to burn in the crowns of trees or shrubs. Fuels are nearly always abundant, and fires ignite and spread when weather patterns produce sufficiently dry conditions. In wetter ecosystems, such as the coastal forests of the Pacific Northwest, this requires a prolonged drought. In others, such as the highly flammable chaparral shrublands of Southern California, a hot, dry wind may be sufficient. In high-severity regimes, attempts to change fire behavior through fuels management are generally ineffective, because fuel regeneration can be extremely rapid and fire behavior is determined primarily by extreme weather conditions.[34]

Mixed-severity regimes combine the features of these two regimes and have both high- and low-severity fires at a variety of frequencies. They are common in midelevation forests, where topography creates a mosaic of tree species, densities, and moisture levels.[35] Fuel abundance and fuel moisture, fluctuating climate patterns, and short-term weather all play a role in the timing and behavior of fires (see figure 1).

It is far beyond the scope of this book to describe in suitable detail the multitude of fire regimes that make up the western United States. In any case, several authors have already done so. The seminal work is Henry Wright and Arthur Bailey's *Fire Ecology: United States and Southern Canada*.[36] An earlier volume edited by T. T. Kozlowski and C. E. Ahlgren, *Fire and Ecosystems*, is also highly informative.[37] More recently, James Agee has written an excellent treatment of the fire ecology of forests of the Pacific Northwest.[38] While they do not delve into specific fire regimes in any great depth, Stephen Pyne and colleagues offer an extremely thorough description of the principles and variables of fire behavior.[39]

Figure 1. A prescribed fire moves across a landscape with a mixed-severity fire regime (juniper shrub and grassland on the 6666 Ranch in central Texas). Photo by Guy McPherson.

Before moving on to other topics, however, we do wish to provide a sense of the complicated situation facing anyone who would seek a national solution to wildland fire issues. In place of a more thorough treatment, we offer case studies of three western ecosystems with vastly different fuels, weather, fire regimes, and management consequences. Remember, however, that the variability among sites, seasons, or years in a given ecosystem can be even more striking than the differences among defined ecosystems. This spatial and temporal complexity is a theme to which we return several times in this book. We see the three case studies here as representing points along a multidimensional continuum of factors that affect fire behavior and fire ecology.

CASE STUDY: FIRE IN PONDEROSA
PINE FORESTS

Forests of ponderosa pine *(Pinus ponderosa)* cover vast swaths of the western United States.[40] The species dominates some 2.7 million acres of the West and forms virtually monotypic stands in many areas. The relatively continuous ponderosa pine forest that stretches along the Mogollon Rim between northern Arizona and central New Mexico is reportedly the largest ponderosa pine forest in the world and forms the only major commercial forest in the southwestern United States.[41] The ecology and fire history of the species are among the most studied in the world.

Ponderosa pine forest varies in density, from very open, parklike stands to dense thickets. This variation in stem density is partially attributed to human-induced changes in fire regimes; fires have been virtually absent in many ponderosa pine forests since the late 1880s.[42] Accounts by some early explorers and settlers describe more open canopies with understories of herbs and shrubs. Understory species vary widely across western North America.[43]

Ponderosa pines have an impressive set of adaptations that help them tolerate fire. Adults trees have thick, plated bark; the spaces between these plates dissipate heat and protect the cambium. The relatively open forest canopy, where it still exists, also allows heat to dissipate and minimizes crown scorching. Buds are encased in small bundles of needles that insulate and protect them from heat damage. Trees are deep-rooted, even at an early age, which also helps them survive fire.[44]

In addition to having considerable resistance to fire, ponderosa pine is well adapted to reproducing after a fire. Although its seeds are large and not particularly well dispersed by wind, fires in ponderosa pine forests tend to be relatively patchy, which helps ensure an adequate supply of seeds from healthy adult trees near recently burned areas. Patterns of reproduction appear to depend largely on soil moisture. Seedlings establish readily in more mesic (that is, less arid) forests but can also recruit in drier areas during periods of above-average precipitation.[45] Once seedlings become established, their growth in terms of both height and diameter is so rapid that trees became large and fire-resistant relatively quickly, although intense crown fires can kill adult trees.[46] Seedlings themselves are moderately susceptible to fire, but mortality depends strongly on fire intensity.[47] This combination of episodic recruitment patterns, rapid growth, and high resistance to fire makes ponderosa pine very well suited to surviving low-intensity surface fires and reestablishing populations in the wake of larger crown fires.

Other species in ponderosa pine forests also have a close relationship with fire. Gambel oak *(Quercus gambelii)* is a dense shrub commonly found beneath stands of ponderosa pine in the central and southern Rocky Mountains. These oaks are relatively resistant to low-intensity surface fires, and, because the species is clonal, individuals damaged by fire often resprout vigorously.[48] As a result, Gambel oak may dominate stands in more arid areas within the first several years after a fire. New Mexico locust responds somewhat similarly. Stems are easily killed by fires, but rapid resprouting and recruitment (which is more successful after seeds have been scarified by fire) allows the species to become locally dominant. These patches of locust usually give way to conifers within fifteen to twenty years. On wetter sites, species such as Douglas-fir *(Pseudotsuga menziesii)*, southwestern white pine *(Pinus strobiformis)*, and quaking aspen *(Populus tremuloides)* coexist more commonly with ponderosa pine. Douglas-fir is thick barked and therefore resistant to all but intense, stand-replacing fires. It reestablishes through wind-borne seeds and grows rapidly, and individuals may persist for several centuries.[49] Southwestern white pine is moderately resistant to fire as a mature tree, but seedlings are highly susceptible to even low-intensity fires.[50] Quaking aspen rarely burns at all when it occurs in pure stands, because aspen leaves and standing trees have a high moisture content.[51] Where aspens are scattered among ponderosa pine forests, however, they may be subjected to more frequent fires. In even relatively low-intensity fires, the aboveground portions of the trees are usually killed.[52] But quaking aspen is the archetypical fire-adapted species in these systems: it resprouts vigorously and grows quickly after a fire, so that it appears on first inspection to be the only tree present. It is also highly shade-intolerant, however, and postfire stands of quaking aspen generally give way to conifers within thirty to sixty years.[53]

These adaptations to fire tell us a great deal about the fire regimes of a given system. Indeed, fire and plant species coexist in a synergistic relationship. The unique characteristics of plants (especially the kinds of fuels they create) shape fire regimes, which in turn favor certain plant adaptations for tolerating fire. Fire and plants are thus inseparable in ponderosa pine and many other ecosystems, and changes in one can have a great impact on the other. Despite a long history of study (or perhaps because of it), considerable debate still surrounds the fire history and ecology of ponderosa pine forests. For many years, scientific consensus has held that, before European settlement, ponderosa pine forests were characterized by frequent low-severity surface

Figure 2. A low-intensity prescribed fire burns in a ponderosa pine forest on the Mogollon Rim in northern Arizona. Photo by Guy McPherson.

fires (see figure 2). Nearly all fires were ignited by lightning; Native Americans may have started a few fires.[54] Fire-return intervals varied in different geographic regions, from about four to thirty-five years.[55] Newer evidence suggests that this long-held understanding of fire in ponderosa pine forests is not entirely accurate. It remains likely that this fire regime predominated on most sites, but some important exceptions are worth noting.

Since at least 2001, considerable new evidence has pointed to the historical occurrence of infrequent, high-intensity crown fires in some ponderosa pine forests.[56] It may be that a cycle of frequent surface fires influenced most stands but that infrequent stand-replacing fires occasionally interrupted that dominant cycle.

A combination of variables related to weather, climate, and stand structure help explain these overlapping cycles.

Fire frequency and intensity are strongly controlled by fuel density and cover in most ponderosa pine forests. These forests are generally found in regions with annual summer droughts followed by dry thunderstorms, so that weather and fuel moisture conditions are favorable for fire nearly every year. Fires occur every few years, or as soon as fuel loads build up to the necessary threshold. Historically, a positive feedback cycle was probably maintained on many sites, in which frequent low-intensity fires maintained relatively open, parklike stands of widely scattered pine trees. However, denser stands probably became established during periods of above-average precipitation and on relatively mesic sites.[57] As a result, in a regime more commonly dominated by frequent, low-intensity fires, occasional high-intensity fires could result from even modest variations in precipitation over relatively short periods of time. For example, a decadelong period of above-average precipitation would cause rapid recruitment and growth, as noted earlier, while also acting to reduce fire occurrence and spread. Fuel loads build rapidly in such situations. A subsequent decadelong period of below-average precipitation would lower fuel moisture and could easily trigger widespread severe, stand-replacing fires. In general, however, large stand-replacing fires were probably rare until the last century or so, when humans largely eliminated low-intensity fires. As we discuss in later chapters, ponderosa pine fire ecology has played a disproportionately large role in recent fire policy and management decisions, and a desire to return these forests to a low-intensity regime has played a major role in reshaping western forests.

CASE STUDY: FIRE IN
MIXED-CONIFER FORESTS

The mixed-conifer forest type of the western United States is found throughout the Pacific states, particularly at the middle elevations of the Cascade and Sierra Nevada ranges. It also occurs in the northern and central Rocky Mountains and the southwestern mountain ranges.[58] Common species include Douglas-fir, sugar pine *(Pinus lambertiana)*, ponderosa pine, Jeffrey pine *(Pinus jeffreyi)*, incense-cedar *(Libocedrus decurrens)*, grand fir *(Abies grandis)*, and white fir *(Abies concolor)*.[59] In the southwestern forests, southwestern white pine and quaking aspen also are common.[60] Canopy density varies among mixed-conifer forests, depending on moisture, elevation, and aspect.[61] Canopy cover and species composition are influenced by other site conditions as well, including soils, temperature, and fire and other natural disturbances.[62] On moist sites (generally, at higher elevations and on north- and east-facing slopes), fir is more common than pine, and forests are normally mature and dense, with a tightly closed canopy and shrubby undergrowth. Drier forests have more pine, a more open canopy, and less undergrowth.[63]

Mixed-conifer forests are characterized by mixed-severity fire regimes, and the relatively complex regimes that create and maintain these forests produce a rich mosaic of trees across the landscape. Before European settlement, fire intensity probably varied considerably even within a single fire, so that some areas did not burn at all, other areas burned as low-intensity surface fires, and some patches burned as high-intensity crown fires. This variety of intensities is undoubtedly responsible for the presence of tree species with very different tolerances to fire. There is little question that fire was a recurrent phenomenon and an integral component of these forests before suppression became common.

Fires are more frequent and less intense in the drier mixed-conifer forests, where pine needles provide abundant fine fuels and all fuels are drier. The wetter mixed-conifer forests, conversely, experience something closer to a high-severity regime—fires are infrequent but intense, and both surface and crown fires occur.[64] In earlier literature the distinction between these two regimes was unclear, leading different scientists to make wildly varying estimates of fire-return intervals. Regimes of more frequent fires tend to create stands of more fire-tolerant species, particularly ponderosa pine, while regimes of less frequent fires favor more shade-tolerant species such as firs and incense-cedar.[65] Tree-ring analyses from southwestern mixed-conifer stands suggest that fires historically occurred at three- to twenty-year intervals.[66] However, in the moister forests of the Pacific Coast, fire scientists Wright and Bailey suggest, intervals ranged from forty to five hundred years, depending on the site and associated species. Fires in both areas tend to occur primarily during major regional droughts and during the dormant season when fuels are driest, and intervals in both have been lengthened considerably by fire suppression.[67]

CASE STUDY: FIRE IN COASTAL DOUGLAS-FIR FORESTS

Douglas-fir is very common in western North America, with a range extending from northern British Columbia to near Mexico City and from the Pacific Coast to the eastern Rocky Mountains. In terms of timber, no North American tree is more significant and productive. Remaining old-growth forests on the northwest coast are legendary for their biological diversity, as well as for the size and growth rate of Douglas-fir trees. These forests harbor some of the nation's most celebrated and contentious rare species, including many species of migratory fish and the northern spotted owl *(Strix occidentalis)*. The ecological communities associated with Douglas-fir vary greatly, as do their fire regimes. These communities are classified according to geography (interior or coastal) and associated species: for example, cedar-hemlock communities in the most mesic, northern climates, quaking aspen and lodgepole pine in interior continental climates, and ponderosa pine in the drier ranges of the southwestern United States.[68] They overlap somewhat with the mixed-conifer forests just discussed, in that these are often dominated by Douglas-fir in the northwestern states. In this example, we focus on the coastal Douglas-fir ecosystems that predominate in the Pacific Northwest of the United States on the western flanks of the Cascade and Sierra Nevada mountains.

In general it is high-intensity, stand-replacing fires, with return intervals of 250 to 700 years, that produce these coastal Douglas-fir forests.[69] Exceptionally deep or long droughts are required for fires in coastal areas, whereas somewhat more frequent but smaller fires, or periodic surface fires, characterize the Cascade Range farther inland.[70] While Douglas-fir is common throughout much of this area, it is generally found in forests dominated by other species. Western hemlock *(Tsuga heterophylla)* and Pacific silver fir *(Abies amabilis)* forests often include Douglas-fir stands. As noted earlier, adult Douglas-fir trees are highly resistant to surface fire, in part because of their very thick bark and fire-resistant seeds. Interestingly, not all associated species are equally resistant to fire.[71]

In the humid, high-elevation silver fir forests, Douglas-fir and silver fir are often joined by noble fir *(Abies procera)*, western red-cedar *(Thuja plicata)*, and western white pine *(Pinus monticola)*. None of these species are as resistant to fire as Douglas-fir, and Pacific silver fir is killed very easily by fire. It has thin bark and shallow roots, and Wright and Bailey suggest that it may require seven hundred to eight hundred years to become reestablished in a burned area, As a result, more frequent large fires limit its distribution.[72] Fire scientist James Agee, however, estimates that its fire-return intervals are between three hundred and six hundred years at higher elevations and between one hundred and three hundred years in lower, drier forests. In Pacific silver fir forests, it is likely that nearly all fires were historically large, intense, and fueled by extreme drought.[73]

The lower, drier western hemlock forests likely experienced a mixed-severity fire regime before Anglo settlement. Low-intensity surface fires were frequent but burned over small areas, and very large, intense crown fires were infrequent but played a major role in shaping landscapes and determining species composition.[74] These large fires, like those in silver fir forests, were typically caused by lightning during significant summer droughts, and they were probably regulated more by weather patterns than by fuel accumulation. In western hemlock forests today, Douglas-fir trees typically dominate after logging or large fires, and they will be taken over by more shade-tolerant conifers after a few centuries.

In coastal forests containing Douglas-fir trees, as in other systems, we see the synergistic relationship between fire and fire-adapted plants. In this case, it is the native Douglas-fir that competes by carrying fire; in nearly all northwestern forests, increased dominance of Douglas-fir is associated with an increase in fire activity, which in turn favors it over less fire-resistant species.[75] This has interesting implications for forest management, as Douglas-fir is highly valued commercially and forms the basis of the northwestern logging industry.

The preceding examples illustrate an intriguing contradiction. We do know, within very broad parameters, where fire "belongs" and about how frequently it should occur to maintain biological diversity in many western ecosystems. We also know, again in very broad terms, the conditions under which fires are likely to occur. But the beauty, mystery, and difficulty of fire occurrence, spread, and behavior are in the details of specific sites and environmental conditions. Of course, even if we fully understood all these details, we still would not have the ability to stop large fires, just as we are unable to stop earthquakes, volcanoes, hurricanes, tornadoes, and other natural cataclysms. Nor should we want to. The environmental consequences of disrupting these large-scale processes would be catastrophic, not least for humans themselves.

Because we now recognize that most species native to the western United States evolved in cycles of periodic fire, we can draw at least two important conclusions. First, these native species have developed adaptations to fires that occur at a particular frequency and season and to a particular extent. Second, because fire is a major disruptive force in the ecosystems it affects, the maintenance or reintroduction of past fire regimes—the fire regimes with which species evolved—is likely a key to maintaining high levels of native biodiversity. Indeed, fire cycles are part and parcel of these ecosystems, just as the cycling of water and nutrients is.

Seen from this perspective, fire is an integral component of nearly every western ecosystem.

But as our examples illustrate, contemporary fires are often unlike the fires with which species evolved. Despite the close synergistic relationships between ecosystems and fire regimes, a century of fire suppression, timber management, the introduction of nonnative species, and countless other habitat alterations have changed the mix of species and wildland fuels in many of these systems. Before we move on to a discussion of management solutions—both successful and unsuccessful—for living with fire, we want to discuss a few of the more pervasive human influences on western fire regimes.

TWO

Fanning the Flames

Human Influences on Fire Regimes

Ask residents of the fire-prone American West where wildland fires come from, and you will hear the same answer again and again: one hundred years of fire suppression. This one factor virtually dominates the national debate over changing fire regimes, to the extent that many people consider the debate resolved. The general consensus is that fires are becoming bigger, more frequent, and more severe, and that federal agencies created this situation with their all-encompassing policy of aggressive fire suppression. According to this argument, fire suppression interrupted natural fire cycles, which has led to heavy accumulations of fuels in many forests. President George W. Bush's Healthy Forests Initiative and associated policies are based on this consensus. But what evidence do we actually have that fire regimes are changing and that fire suppression is at fault? If suppression is causing more and bigger fires, is it possible that other forces are also at work?

National data indicate a slow increase in the number of acres burned in wildland fires from 1960 to 2005, although the trend is

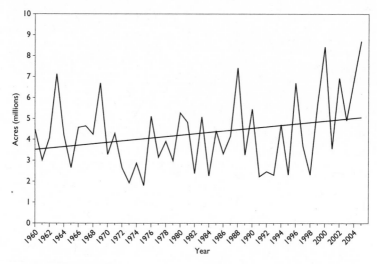

Figure 3. Acres burned per year in the United States (including prescribed and naturally ignited fires), 1960–2005. Data from the National Interagency Fire Center.

obscured somewhat by a great deal of year-to-year variability (figure 3).[1] The number of individual fires, however, shows no clear linear trend during the same time period (figure 4). The number of fires reported has decreased since the late 1970s and early 1980s but is more or less the same as in the 1960s and early 1970s; however, again, there is a great deal of interannual variability.[2] These patterns imply that the average size of fires is increasing and the frequency is not.[3] This might be attributable to managers increasingly letting some lightning-ignited fires burn unchecked. Disturbingly, the annual number of firefighter deaths rose substantially during that time period (figure 5).[4] In the absence of any other clear explanation, this may suggest that at least some fires are being fought more aggressively. Perhaps more fire-

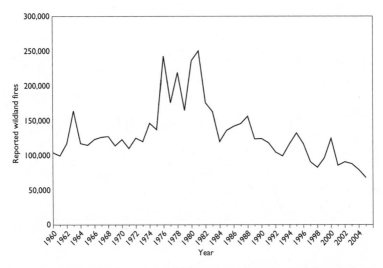

Figure 4. Number of wildland fires reported annually in the United States, 1960–2005. Data from the National Interagency Fire Center.

fighters are being deployed to fight these larger fires, or perhaps more assertive tactics are being used. Regardless, it appears that more firefighters are being placed in harm's way.

If we agree that the average area burned has increased over time, there are many reasons to conclude that the last century of fire suppression is at least partly responsible. In many North American forests, fires have been virtually excluded for a century or more in response to a federal policy of 100 percent suppression that is only now coming to an end. This policy has lengthened fire intervals in many ecosystems characterized by low-severity fire regimes (as in the ponderosa pine example in chapter 1). In some cases the change has been dramatic; for species that evolved with fires occurring every five or ten years, a century of fire suppression

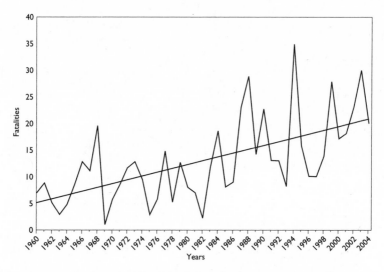

Figure 5. Number of wildland firefighter deaths reported annually in the United States, 1960–2005. Data from the National Interagency Fire Center.

is a very long time. In these ecosystems, fire exclusion has allowed fuels in forest understories to build up to unprecedented levels. Where understory fuels have accumulated, forests are predisposed to high-intensity, stand-replacing crown fires. In such ecosystems, it may be desirable to reduce fuels to protect humans and buildings, as well as fragile soils and species adapted to lower-intensity fires.[5] In later chapters we provide much more detail on the theory and practice of managing fires by reducing fuels.

There are plenty of reasons to think that fire suppression is not the sole cause of large or intense fires, however. We know that large, intense fires were not uncommon, and that much greater areas burned annually, before 1911, when the passage of the Weeks Act heralded the beginning of the era of broad-scale federal fire suppression.[6] For example, the worst fires in the history of western North America, the Great Idaho Fires of 1910, occurred at the very beginning of what is so often called the "century of mismanagement"; in fact, they were a major cause of that fire-suppression policy. The scale of these fires is virtually unheard of today. They burned 5 million acres of federal land alone, compared with a total of 1,585,000 for the 1988 Yellowstone Fires or 462,000 for the Rodeo-Chediski Fire in Arizona in 2002.[7] The Great Idaho Fires were preceded by the Peshtigo Fire of 1871, which charred more than a million acres of the upper Midwest and took hundreds of human lives.[8] Only in Alaska, where fire suppression has never been comprehensive, do fires of this magnitude still occur.[9]

Fire exclusion has likely had little influence on ecosystems where species have evolved with very infrequent fires. In a system with an average fire-return interval of several hundred years, a century of fire suppression is unlikely to have made much of an impact. Furthermore, high-severity regimes, in which fires may be frequent or infrequent, are influenced less by fuel accumulation than by weather and climate. In these cases, any fuel buildup because of fire suppression is unlikely to affect fire frequency, size, or severity. Notwithstanding considerable rhetoric from politicians, scientists, and land managers, fire suppression has not single-handedly created the "fire crisis" described each summer by the media.

Realistically, the contribution of fuel accumulation to the current wildfire situation varies from one site to another and depends on myriad characteristics of ecosystems and land use.[10] The paradigm that sees changing fire regimes as the direct result of fire-suppression-induced fuel buildups is based largely on the study of ponderosa pine forests. In many other ecosystems, suppression policy often plays only a minor role in creating larger and more intense fires, especially when compared to other land-use factors.[11] However, four other anthropogenic causes play major parts in changing fire regimes: global climate change, unrestrained development in the wildland-urban interface, inappropriate logging and grazing practices, and introductions of nonnative species.

CLIMATE CHANGE AND FIRE REGIMES

Weather and climate are different but closely related phenomena. Robert Heinlein describes the difference this way: "Climate is what you expect, weather is what you get."[12] Weather does change from day to day and hour to hour, and land managers find planning for its effects difficult. Our traditional understanding of climate, however, was that it changed slowly, over centuries and millennia. Great shifts caused glaciers to advance and recede, but the vast time scales involved meant that these changes were usually unobservable within a human lifetime. Today, both our understanding of climate and the nature of climate change itself are shifting.

While popular media coverage around the turn of the twenty-first century has often suggested that the jury is still out on global climate change, the scientific consensus on the matter is painfully clear on the basic points. The earth's surface is warming at a rate that appears to be unprecedented over at least the last ten thousand years, and this warming is at least partly the result of human activities. Since the beginning of the industrial revolution, human-caused emissions of greenhouse gases have increased exponentially, and most of the global warming trend over at least the last fifty years can be attributed to these emissions. As a result, the next few centuries or millennia will be marked by significant changes in global climate, the magnitude and duration of which might depend greatly on our actions over the next few years. Globally, we should expect at least three interrelated kinds of changes: an increase in atmospheric carbon dioxide (CO_2) levels, an increase in average surface temperatures and a concomitant rise in sea levels, and more extreme hydrological cycles (including an increase in globally averaged precipitation, more intense local precipitation events, and increased risk of drought in some areas).[13]

Aside from the denials of very few researchers, the scientific uncertainties that do still exist involve either the magnitude of global changes expected under various mitigation scenarios or the nuts-and-bolts details of regional and local effects. While most models of greenhouse gas emissions agree on these basic global trends, it is much more difficult to predict the sometimes contradictory changes that may occur in any given country, ecosystem, or management area. For example, while an increase in average global precipitation is expected, some regions will actually experience more frequent and severe droughts.

Similarly, the changes predicted within the United States are mixed and uncertain—climate change will vary widely across the United States, but climate scientists also have less confidence in predictions they make at this scale than at the global level. In general, warming in the United States is predicted to be greater than the global average, in part because more warming is expected at the higher latitudes. Models differ a great deal in the spatial and temporal patterns of warming they predict. Most major climate-prediction models show an increase in precipitation, especially in winter, in California and the southwestern United States, but results for the rest of the country are less clear. A rise in sea level may cause significant coastal erosion and even coastal flooding by the end of the twenty-first century. It is also likely that the increased occurrence of unusually large storms, seen recently, will continue. As a result of these effects, some regions are likely to experience a rise in plant productivity, while others (most notably the southeastern part of the country) may experience decreased productivity. Scientists also expect significant changes in vegetation distribution.

Fire historian Stephen Pyne writes, "Natural fire is a climatic phenomenon."[14] Climate plays a significant role in determining the distributions of different species, each of which, as we have seen, has its own unique adaptations to fire. It also affects the overall productivity of ecosystems—and thus the production of potential fuels—and the structure and moisture content of vegetation.[15]

It is safe to assume that any changes in temperature and precipitation will make fire management an even more complex and challenging endeavor than it is today. In the western United States, scientists anticipate a northward shift in the geographic range of vegetation types, with some species and even ecosystems likely to disappear completely. Increases in insect and disease outbreaks, changes in forest function, and increases in fire frequency are also expected.[16]

Why should we expect more frequent wildland fires in a world with increased temperatures, precipitation, and CO_2 levels? Stated simply, increased temperatures can be expected to increase the length and severity of fire seasons, as hotter weather makes fuels more combustible for a greater portion of every year. In the western United States in particular, increasing temperatures might be associated with a decrease in average relative humidity, which would exacerbate this effect.[17] High-severity fire regimes will likely be affected greatly by this phenomenon, because fire occurrence is so closely tied to periods of hot, dry weather.[18]

A changing climate will also undoubtedly place great stress on forests. Trees are, of course, unable to move or otherwise adapt quickly to rapidly changing conditions, so high mortality can be expected. Some of this stress and mortality will result directly from changes in temperature and precipitation (remember that precipitation, while increasing on average, may decrease in some areas). Insects and diseases are also major sources of stress to trees—in fact, the effects of their disturbance on forests are much greater than that of fire itself. Hotter, drier conditions are expected to increase these pest populations.[19] All these changes will likely be associated with more frequent and severe fires, because dead and dying trees provide highly combustible fuel.

A doubling of atmospheric CO_2 is generally expected to increase overall plant productivity. We anticipate an associated increase in fire frequency in fire regimes heavily influenced by fuel accumulation. But atmospheric CO_2 does not affect all plants in the same way. For example, controlled experiments show that higher concentrations of CO_2 favor the growth of nonnative annual grasses in the southwestern deserts.[20] As we will see later in the chapter, these grasses create a positive feedback cycle that has already greatly accelerated the fire regime: they supply fine fuels that are crucial to fire spread but that were historically absent from these desert ecosystems. Indeed, under many climate-change scenarios, vegetation across the Southwest is expected to shift and produce more fire-prone shrub- and grassland.[21] This shift in vegetation may exacerbate an already precarious situation for fire managers, residents of desert exurban areas and even cities, and native desert plants and wildlife.

Finally, the more extreme weather patterns predicted under some climate models would also affect fire regimes both directly and indirectly. In high-severity fire regimes, where fire is virtually always weather-driven, rapidly shifting patterns of extreme weather could greatly increase its frequency.[22] An increase in severe storms would likely lead to more lightning ignitions, especially in the stressed and dry forests we have described here. Severe storms often lead to windblow, a phenomenon in which most or all trees in a large area are felled by high winds. This also adds highly combustible dead fuels to forests, so we can also expect low-severity regimes to be affected.

Recent research suggests that we are already seeing these trends in action. Scientists analyzing wildfire activity have found a sudden increase in the frequency and duration of large wildfires, beginning in the mid-1980s. As anticipated, the wildfire season is also becoming longer, with the first wildfires occurring earlier in the spring and the last ones burning later into the fall. Such changes are strongly associated with regional and subregional changes in climate and hydrology. For example, these changes are especially noticeable in the northern Rockies, where high-severity fire regimes dominate and land use changes have had a relatively minor impact on fire risk.[23] As climatic changes continue to develop, we can expect more of the same: longer fire seasons and longer-lasting and more frequent large fires.

Global climate change is real, and it is currently observable. Scientists have developed different climate change scenarios based on different assumptions about future greenhouse gas emissions. Even if greenhouse gas and aerosol emissions had been permanently fixed at the levels emitted during the year 2000, we could still expect a global average of 0.2°C of warming over the following decades. More realistic models assume a continuing increase in greenhouse gas emissions. Of these, the most conservative estimate suggests an increase of about 1.8°C between 2029 and 2099. Models based on larger predicted increases in emissions suggest an increase of 4°C over the same period.[24]

The potential extent and variety of climate change effects can be mind-boggling. Some of the predicted effects appear to be contradictory or even mutually exclusive, while others are so complex as to be virtually unfathomable. From centuries of experience, however, we know that fire tends to thrive in a changing landscape.[25] Changes in climate, weather, vegetation, and land-use patterns are typically accompanied by a change in fire regimes, and all these kinds of changes can be expected in the years to come.

In the western United States in particular, demographic changes are driving massive transformations in land use. Urban expansion, extractive management, and the increasing spread of some species out of their native environments are changing ecosystems and fire regimes in unexpected and even dangerous ways.

THE PAVING OF THE WEST

On October 25, 2003, a hunter lost in the Cleveland National Forest lit a signal fire in central San Diego County in Southern California. The fire started in California chaparral vegetation—which is characterized by dense, highly flammable shrubs—as the hot, dry Santa Ana winds blew from the desert situated to the east. Overnight, it spread across 100,000 acres and killed 13 people living near the town of Lakeside. It was quickly contained, but the short-lived Cedar Fire was the second largest fire in California's history. In about twelve days, it burned more than 275,000 acres, caused 17 deaths, injured 104 firefighters, and destroyed 2,400 structures. Many residents of the area blamed the damage on a slow and insufficient response by firefighters, but chances are good that nothing firefighters could have done would have affected the fire's course. The Cedar Fire was one of fifteen fires that started in the area in late October that year, and all those fires combined burned 721,791 acres and 3,640 homes.[26]

California chaparral is the epitome of a fire-prone environment. Vegetation here is shrubby, with abundant but loosely arranged small fuels; chamise *(Adenostoma fasciculatum)* and various species of manzanita *(Arctostaphylos spp.)* are the most common shrubs. High-severity fires are not only typical but also frequent. The dominant plant species grow exceptionally fast, so fuels rebuild quickly after a fire. Fire behavior, as in other high-severity regimes, is primarily determined by weather. To make matters worse (from the perspective, at least, of human residents in the chaparral), many species of plants contain highly flammable oils. Because of these abundant, volatile fuels, and because extreme fire weather frequently occurs here, fires arise often. Fires are driven by the extremely hot, dry Santa Ana winds, which are capable of spreading fire even through very young stands. Large fires of more than 5,000 acres generally occur every twenty to forty years.[27]

There is a widespread belief among residents, policy makers, and even some scientists that California chaparral, like ponderosa pine forests, is experiencing increased fire activity because of aggressive fire suppression. They suggest that forcibly creating a regime of small, frequent fires will prevent very large, intense fires. This application of a low-severity paradigm to a high-severity fire regime is totally misguided. Attempts to change fire behavior either by reducing fuels or by suppressing fire are largely futile, and fire suppression efforts have never been effective in chaparral. Fires in this ecosystem have not increased in size, season, or severity since at least the late 1800s. The frequency of fires has increased in recent years, however, and this increase is the direct result of human population growth in the region. As Southern California's suburbs expand into the surrounding chaparral, human actions increasingly join the Santa Ana winds in driving this tinderbox's fire regime.[28]

Population growth in Southern California is, of course, nothing new. Since World War II, the region has doubled its population every decade.[29] The trend has been repeated in several other parts of the western United States, most recently in the southwestern states. Also since the mid-1940s, the population has been moving out from urban areas into the suburban, exurban, and rural areas.[30] It is in these nonurban areas, where large numbers of people live in or adjacent to fire-prone wildlands, that demographic changes are affecting wildland fire regimes. These areas are often referred to as the wildland-urban interface, and they affect every facet of fire management today.

In any such area, fire frequency tends to be greater because there are more humans to ignite fires, whether intentionally or unintentionally. Today the majority of wildland fires are human caused.[31] Anthropogenic ignitions tend to occur close to roads, campgrounds, picnic areas, and urban areas.[32] Increased human ignitions are the primary cause of the accelerated fire cycle in Southern California chaparral ecosystems.

Similarly, development in the wildland-urban interface can undermine attempts to manage fire. In the classic wildland-urban interface, suburban sprawl brings densely populated developments into contact with wildlands. In these areas, even a small fire may threaten many houses. This problem is exacerbated when residents fail to clear potential fuels from around their homes.

In some cases, the subdivisions themselves may be even more fire-prone than the surrounding ecosystems.[33] For example, when the infamous 2002 Cerro Grande Fire (which threatened the Los Alamos National Laboratory) entered the town of Cerro Grande, New Mexico, it was burning at ground level as a relatively calm surface fire. When it hit the edge of town, however, the fire grew in intensity, and its behavior became more extreme. In residential areas, fuels, including woodpiles, flammable shrubs, a continuous ponderosa pine canopy, heavy pine-needle beds, and even the homes themselves, ignited, and the fire spread from structure to structure. Some homes did not ignite, simply because residents had taken the most basic precaution of sweeping pine needles away from the bases of their houses, where even a low-intensity fire could ignite wood siding and cause the entire house to burn. When Jack Cohen, a research scientist with the USDA Forest Service, examined patterns of home destruction after the fire, he found many instances in which homes were totally destroyed while the trees next to the houses were only scorched. Where trees were completely consumed by fire, they appeared to have been ignited by burning houses rather than by the low-intensity surface fire. This clearly indicated that the area's excess fuels lay not in the forests but in and around the houses themselves. He concluded that heavy residential fuels had contributed significantly to the town's damage.[34]

Areas where buildings are scattered present their own challenges to fire managers. When homes are isolated, they tend to be difficult to evacuate and protect once a fire starts. They may also border public lands where managers are attempting to reintroduce fire or where fire protection is impractical. City fire protection may not be available, so homeowners in such areas often depend on Forest Service firefighters or a local volunteer fire department for protection.

The federal government's efforts to protect the wildland-urban interface have been herculean. In 2005, the five federal land-management agencies spent $875.7 million on wildland fire suppression.[35] During the more active 2002 season, the Forest Service alone spent more than $1.2 billion.[36] These expenditures occur even as those same agencies blame the fire crisis on the previous century of fire suppression. Because of the political unacceptability of "losing" properties to fire, protecting private property remains a very high priority in all fire management activities, second only to protecting human life.[37] As we will see, this protection comes at substantial ecological and social costs.

EXTRACTIVE MANAGEMENT

Shortly after the 2002 Ryan Fire burned thirty-eight thousand acres of southern Arizona grassland, a sign appeared near the site of the fire. The sign, reading "Grazing prevents blazing," depicted an obviously distraught cow glaring down at the word *blazing*, the letters of which were aflame (figure 6). This sign, placed by the Elgin/Sonoita Cowbelles, echoed others often posted across the rural West. These signs suggest, of course, that livestock grazing prevents wildland fires and presumably could have prevented the highly contentious $1.2 million Ryan Fire.[38] The Ryan Fire, however, started in a pasture being grazed by cattle, and most of the burned area had been recently grazed.[39]

Logging, like grazing, has occasionally been proposed as a tool for wildfire prevention. For example, representatives of the timber industry suggested that logging could have prevented the

Figure 6.   A southern Arizona road sign advocates fuel reduction
through grazing. Photo by Guy McPherson.

1988 Yellowstone Fires.[40] However, both of these extractive uses
tend to increase fires in some ecosystems, and there is no compelling evidence that they reduce fires in any. Grazing and logging are common, albeit increasingly controversial, practices on
public lands. They are common also on privately owned lands
in the wildland-urban interface, although the expansion of residential development in the rural landscape has led to a decrease
in extractive uses. Because these activities affect such vast areas of
the western United States, especially on public lands, it is important that we have an accurate idea of their impacts on fire regimes.

Fire historian Stephen Pyne has argued, "More than any 'fire
practice' per se, the wholesale introduction of domestic animals

reconfigured the fire regimes of the continent."[41] In the first half of the twentieth century, the idea that livestock grazing reduced fire frequency was popular. Aldo Leopold, a great American conservationist, was a proponent of grazing southwestern grasslands to prevent wildland fires.[42] The reasoning behind this argument seems logical: grazing removes fuel from the landscape, therefore it must reduce the probability of ignition and inhibit the spread of fire. It is true that grazing may initially reduce fire frequency or intensity by removing fine fuels, especially in grasslands, where fine fuels play a large role in carrying fire. But in most forests, herbaceous plants contribute little to the fuel load. Rather, nearly all the available fuel in most forested ecosystems is downed woody material (fallen trees and branches), low-hanging dead branches, and conifer leaves desiccated by hot, dry weather.[43] Even in grasslands, however, it is common for fires to ignite and burn on grazed lands, as happened in the Ryan Fire.

Not all livestock are alike, of course, and not all ecosystems respond the same way to grazing. Nonetheless, some general concepts, principles, and effects are associated with livestock grazing as it affects western ecosystems. First, considerable research demonstrates that cattle prefer grass; sheep eat grasses but prefer other soft, nonwoody plants (forbs, also known as herbaceous dicots); and goats prefer to browse woody plants. We focus primarily on cattle because they are by far the dominant livestock on western rangelands, but there is some evidence (primarily anecdotal) that livestock such as goats that preferentially browse woody materials may reduce fire frequency or intensity in some systems. As a result, goats are increasingly being used to reduce woody fuels in small areas, especially in the wildland-urban interface. As with any livestock, however, goats' hooves can easily damage the fragile biological crusts that help stabilize soils and protect against invasion by nonnative species.[44]

Grazing by cattle does remove fine fuels in the form of grasses. Paradoxically, though, this sometimes leads to the development of an even more flammable landscape. When grasses and other herbaceous plants are removed from an area, other plants move in. This phenomenon has been recognized for at least a century: reporting on the status of grazing lands in New Mexico in 1908, an early range scientist observed, "Stock eat the valuable forage plants and leave the poor ones, thus giving the latter undue advantages in the struggle for existence."[45] The same pattern is evident at several scales. For example, cattle eat grass and therefore encourage the growth of woody plants, but they also prefer some grasses to others. Long-term grazing by cattle leads to an ecosystem's dominance by woody plants and coarse, unpalatable herbs.

Cattle favor the recruitment and growth of the woodier plants not only through direct grazing but also by making the ecosystem in general less hospitable to grass growth. Livestock remove vegetation, which causes the soil to warm. Warm soil favors the germination and emergence of woody plants into areas formerly dominated by grasses.[46] Grazing also removes organic matter, thereby further favoring the establishment and survival of woody plants at the expense of grasses. The hooves of livestock compact soil, decreasing the infiltration of water and increasing erosion; these changes, too, tend to favor woody plants. Fire is another major instrument behind the shift from grasses to woody plants: the initial reduction in fire frequency that results from grazing accommodates woody plants, which are more susceptible to fire than grasses.[47] In some grasslands, particularly in the southwestern United States, grazing and the associated reduction in fire frequency can actually convert grasslands into shrublands.[48] This transition creates serious negative consequences for both biological diversity and livestock ranchers themselves.

In some ecosystems, livestock grazing can actually increase fire severity simply by decreasing its frequency. In many dry ponderosa pine forests, stand-replacing fires have become more common at least in part because livestock grazing (along with fire suppression) helped eliminate the previous regime of frequent, low-intensity surface fires. In many areas, grazing has been encouraged in forests because it reduced fire frequency. But this reduction has also been largely responsible for the high density of ponderosa pine trees in many areas, and these forests have become the "poster child" for high-intensity, stand-replacing fires that now threaten lives and property.[49] Therefore, in ecosystems where the "century of mismanagement" paradigm helps explain increased fire severity, livestock grazing must also be recognized as a crucial part of that mismanagement.

In an interesting side note, the conversion of grasslands to shrublands through livestock grazing has been proposed as a partial solution to the patterns of global climate change. Some researchers have suggested that the increase in woody plants might function as a potential long-term sink for atmospheric carbon, to help mitigate the enormous quantity of atmospheric carbon dioxide produced within the United States.[50] In terms of greenhouse gas production, unfortunately, the carbon and methane produced by livestock quickly overwhelm this "carbon credit." Grazed sites store less soil carbon than ungrazed sites, and this difference approximately balances the carbon stored by woody plants that have encroached into grazed areas.[51] Also, according to the U.S. Environmental Protection Agency, livestock account for about 20 percent of anthropogenic methane production via flatulence and belching; as a greenhouse gas, methane is about twenty times more potent than carbon dioxide.[52]

It is also important to note, at least in passing, the plethora of negative effects livestock grazing is known to have on western grasslands and forestlands. As we have already discussed, ranching is common on public lands, and although the number of livestock on public lands has steadily declined in recent years, the conflict between livestock grazing and the conservation of biological diversity continues to spark considerable debate.[53] The primary negative ecological effects of common ranching practices include the eradication of top predators; the fragmentation of natural landscapes by fencing, roads, and other features; the accelerated spread of nonnative species (as well as the use of herbicides to control them later on); and the draining and contamination of wetlands and riparian areas. Each of these, alone or in combination with the other factors, engenders a loss of biological diversity in wildlands.[54] They may also exacerbate the less desirable economic, social, or ecological effects of fires by leaving ecosystems fragile and impoverished.[55]

Livestock grazing, then, can reduce fire frequency in some ecosystems, but at a significant cost: increased fire severity, reduced biological diversity, and fundamental and possibly permanent changes to the natural landscape. In other systems, including many relatively mesic forests, it has little or no impact on fire but has generally negative ecological impacts. How does logging, another key extractive use in western forests, compare? Does it cause or prevent wildland fires . . . or neither?

Like grazing, logging removes flammable materials, so it might also be expected to reduce fire activity. Like grazing, it generally does not; the ecosystem response to the removal of fuels is complex. In most cases, commercial logging operations actually increase the availability of some classes of highly flammable plant material. These operations leave behind dry, dead fuels in the form of slash (the branches and other residue left on a forest floor after timber is removed) and damaged trees. Wright and Bailey describe a just-logged cedar-hemlock forest that contained more than one hundred tons of fuels per acre in the form of logging slash.[56] Logging also creates even-aged stands and forest fragmentation. This in turn leads to more frequent and intense fires, and it exacerbates the impact fires have on ecosystems by damaging soils, disturbing wildlife, and changing vegetation structure.[57] Logging can also change fire regimes by increasing the dominance of different species. In forests that contain Douglas-fir, logging increases the presence of that highly flammable and fire-tolerant species and thereby tends to increase fire frequency.[58]

While traditional timber harvesting's tendency to accelerate fire activity is widely recognized, two other forms of logging are still often used to prevent wildland fire or minimize its severity. The first is the removal of smaller-diameter fuels from dense forests that land managers consider "overstocked." We address these mechanical fuel-reduction projects, popularized by the Bush administration's Healthy Forests Initiative, in greater depth in later chapters. For now, it will suffice to point out that the scientific rationale for these projects is based almost entirely on research conducted in southwestern ponderosa pine forests characterized by relatively frequent, low-intensity surface fires, and that mechanical fuel reduction can have little or no impact on fire activity in ecosystems typified by weather-driven, high-severity fire regimes.

The second modern form of logging for fire prevention is salvage logging, which consists of removing timber from recently burned forests or from forests damaged by insects, winds, or other natural disturbances. Like grazing and traditional logging, salvage logging might appear to reduce fire hazard, and it does remove the highly flammable dead vegetation that all fires leave behind. Some land managers argue that forests will not recover after severe fires unless dead matter is cleared. Unfortunately, this is yet another misunderstanding based on an oversimplified picture of fire. A study of postfire logging after the 2002 Biscuit Fire in Oregon shows that salvage logging significantly hampered regeneration by disturbing soil and burying or trampling new seedlings. Furthermore, the salvage logging operation, like other forms of logging, produced large amounts of slash and left it on the forest floor. On unlogged sites revegetation was rapid and successful, making seeding virtually unnecessary. The slash left behind after salvage logging may actually increase the flammability of burned forests; at the very least, it clearly fails in its intent to reduce fuels and encourage reforestation.[59]

Salvage logging also causes substantial ecological damage and may, in some cases, cause more damage than the original fire or other natural disturbance.[60] Salvage operations disrupt and destroy vegetation, soils, water, and wildlife. Many species in fire-prone forests survive fires, whether by leaving the area as it burns (in the case of animal species), enduring the fire, or recolonizing the burned area after the fire. Dead trees and branches provide valuable wildlife habitat, and forest regrowth and regeneration begin quickly on the newly nutrient-rich forest floor. Contrary to appearances, a recently burned forest is highly diverse and teeming with life. At the same time, soils are unstable and tend to compact or erode easily. Such a landscape is highly susceptible to damage by heavy logging equipment.[61]

ALIEN INVADERS

In addition to global climate change, unrestrained development in the wildland-urban interface, and inappropriate logging and grazing practices, contemporary land managers are increasingly concerned with nonnative species, those plants and animals that are transported from their native ecosystems and manage to become established in new ones. They may be intentionally introduced for food, fodder, landscaping, or other purposes, or humans or animals may carry them to a new location by accident. Some of these species are so successful in their new environments that they are described as "invasive": they out-compete native species and can eventually undermine local diversity. Invasions by nonnative species are a major cause of extinctions around the world.[62] Fire ecologists don't yet fully understand how nonnative species affect fire regimes, or vice versa, but the following example of fire in western deserts shows how complex and involved that relationship can be. This example is drawn from the almost infinite number of ecosystems undergoing rapid change, and is intended to illustrate rather than predict or comprehensively explain the phenomenon.

Historically, North America's deserts rarely if ever burned; the lack of fine fuel prevented recurrent fires in deserts with less than seven inches of annual precipitation. All other ecosystems on the continent burned periodically before European settlement.[63] The Mojave and Sonoran deserts of the southwestern United States, as well as the "cold desert" of the Great Basin, have typically consisted mainly of low-growing, and often relatively incombustible, shrubs and succulents, and the lack of groundcover between these plants historically inhibited the spread of fire. On the rare occasions that fires did occur, they virtually never became large or intense. As a result, desert plants are generally poorly adapted to survive fire, even low-intensity fires.[64] Since the mid- to late 1970s, however, fire frequency in the deserts of the southwestern United States has increased sharply.[65] The primary cause of these fires is almost certainly a large-scale invasion of nonnative annual grasses. These grasses provide the missing fine fuels to carry fire in these ecosystems.

The most common nonnative grasses in western deserts are various species of *Bromus* (especially *B. rubens*) and *Schismus* (especially *S. barbatus*). Both genera include annual grasses native to Europe, and their original introduction, through livestock grazing, appears to have been unintentional.[66] Initially, the rapid establishment and spread of these species was probably primarily the result of a long period of above-normal rainfall between 1976 and 1983; for example, in the Mojave Desert at Palm Springs, California, annual precipitation was greater than normal for five of these seven years and reached 299 percent of the annual mean in 1979–1980.[67] However, the role these species play in changing fire regimes also helps explain their success.

When annual grasses are established in a desert ecosystem where groundcover has previously been scarce, they create a positive feedback cycle of fuel and fire.[68] The newly established grasses provide the fine fuel needed to start a fire and the continuity of fuel needed to carry it. The annual species mentioned earlier are particularly successful at carrying fire because they decompose more slowly than native grasses. This quickly leads to

Figure 7. A saguaro cactus burned by a rare (but increasingly common) desert fire near the Boyce Thompson Southwest Desert Arboretum in central Arizona. Photo by Guy McPherson.

the accumulation of large amounts of highly flammable fine fuel.[69] In general, these nonnative species come from fire-prone Mediterranean ecosystems, and they recover more quickly from fire than do native species, which evolved with little or no fire.[70] Therefore, increased fire frequency, intensity, and extent result from the predominance of nonnative grasses and, in turn, create conditions favorable to their continued success.

This fuel-fire feedback cycle is highly effective at spreading nonnative annual grasses and is expected to substantially change the physical and biological nature of desert ecosystems. What kinds of long-term effects can be expected? With few exceptions, fires are much more destructive to woody species and succulents than to grasses, which generally lose only one year of growth.[71] Most native species also lack adaptations to fire that would allow them to recolonize an area rapidly after fire. A study of south-western creosote bush *(Larrea tridentata)* scrub, for example, suggests that postfire recolonization by native perennials is so slow that this would likely require hundreds of years.[72] Another particularly compelling example is the saguaro cactus *(Carnegiea gigantea)*, the signature plant of the Sonoran Desert. It evolved in the absence of recurrent fires, so it has no evolutionary strategy to deal with fires. This most majestic of desert plants is readily killed by fires of any intensity (see figure 7).

As we begin to discuss past, present, and future attempts to "solve the fire problem," we should keep in mind the multitude of human activities that shape modern fire regimes. Where fire regimes are changing, these changes can be traced not to a single cause but to a number of interrelated causes. Most or all of these causes are related to changing and expanding human activities. The buildup of fuels resulting from fire suppression often plays a minor role in the spread of economically destructive fires compared to the roles of weather and poorly devised development in the wildland-urban interface.[73] In many cases we can and should try to limit or mitigate human activities. We can reduce our carbon emissions, protect wilderness areas, and try to control nonnative species, knowing that doing so may help restore some ecosystems to the fire regimes under which they evolved. Since we cannot hope to stop affecting fire regimes, however, we cannot hope to restore all wildlands to their historical or prehistorical regimes. Neither can we expect to create fire regimes that fully accommodate human wishes.

THREE

The Failed State of Fire Suppression

In a compelling and insightful essay on collaborative management, the environmental law scholar Bradley Karkkainen offers three propositions about ecosystems that, he says, "I hope are uncontroversial." The first states that "ecosystems are complex dynamic systems." We aimed in the first two chapters of this book to show that this certainly holds true in the case of fire regimes, and that, as Karkkainen continues, they are "composed of many mutually interdependent parts operating in dynamic, co-evolutionary trajectories." The functions of such a system are inevitably difficult to analyze or predict.

A second proposition by Karkkainen is that most ecosystems have now been influenced or even dominated by humans. In the previous chapter, we presented some of the ways human activities have changed fire regimes over the past few centuries. It would be no exaggeration to say that humans dominate fire regimes. Most human efforts to control fire regimes have had unintended effects, and many activities not expected to affect fire at all (for example, our nation's heavy dependence on greenhouse-gas-producing fossil fuels) have actually changed fire regimes rather drastically.

Karkkainen's third proposition is that "the ecosystem context matters in environmental decision-making." Just as no two ecosystems are alike, "the environmental consequences of our actions may also vary widely, depending upon the particular ecosystem context in which the action occurs." If we as a nation hope to manage fire and live with it productively, we must explicitly account for this fact. Ultimately, we must find a way to cope with this variability and attend to the unique characteristics of each ecosystem and locality. Previous chapters have touched on this issue, but we now turn to it in earnest with a depiction of what happens when humans forget how important local context can be.[1]

On June 23, 2002, two large human-caused fires merged in east-central Arizona to form the Rodeo-Chediski Fire, a fire that would ultimately burn across 462,000 acres and destroy 462 structures.[2] The Rodeo-Chediski Fire, the largest in Arizona's history, was in many ways emblematic of increasing public concern over fire management in the wildland-urban interface. The Chediski Fire was started on national forest land by a lost hiker, and the Rodeo Fire was started by a part-time firefighter on the Fort Apache Reservation.[3] Roughly two-thirds of the total area that burned consisted of tribal and private lands. Millions of dollars' worth of timber was destroyed on tribal lands. The total cost of suppression was approximately $153 million. The combined fires threatened thirty communities and subdivisions, and authorities evacuated thirty thousand residents.[4] Observers were made keenly aware that fire-management practices on federal lands often affect adjoining private land as well. This includes not only the properties with typical summer cabins nestled in flammable forests but also fairly large, well-established towns in the wildland-urban interface.

Unlike the Yellowstone Fire, the Rodeo-Chediski Fire took place mainly in an ecosystem dominated by ponderosa pine that developed with a low-severity fire regime. Estimates of the fire-return interval for this ecosystem range from 2 to 25 years, to 22 to 308 years, depending on how tree-ring data are interpreted.[5] In the relatively dry ponderosa pine forests of the southwestern United States, the exclusion of fire for nearly a century has contributed to the development of dense forests that are very different from the open, parklike stands described by early European settlers. These forests have become more susceptible to intense crown fires, whereas more open stands tend to have grassy forest floors that carry low-intensity surface fires. Where fires and mechanical thinning have reduced the amount of fuel in these forests, fire behavior and intensity are generally less extreme.[6] The Rodeo-Chediski Fire provides a perfect example of this. Areas treated with prescribed fire within the previous nine years were burned much less severely than untreated areas or those that had been treated less recently.[7]

We have seen that fire suppression is not the only cause of dense forests; climate change, livestock grazing, and commercial logging all bear some responsibility for altered ponderosa pine forests. The Rodeo-Chediski Fire burned through forests that had been heavily logged, and it occurred during an unusually severe drought.

For much of the last century, our understanding of fire was dominated by the assumption that it was both possible and necessary to completely exclude fire from western ecosystems. Fire policy and management strategies from 1910 to the 1970s were driven by this assumption, and as fire-suppression techniques became progressively more effective, the result was ecological change on an astonishing scale. Although fire is now seen as an important component of most wildland ecosystems, the policies and strategies derived from the outdated suppression paradigm are still widely used. The negative environmental impacts of a century of broad-scale fire suppression continue to shape natural landscapes. Social and political landscapes are also negatively affected by overly aggressive fire suppression, as the costs of this failed policy continue to mount with every new fire season. Perhaps a thorough accounting of these social and ecological costs will help purge the old methods and usher in a new age of intelligent and informed fire management.

FIRE SUPPRESSION: THE GOLDEN YEARS

In the early twenty-first century, wildland firefighters seem to be fighting a losing battle. It is not uncommon for the federal government to spend $1 billion or more on fire-suppression efforts in the course of a fire season, yet we see record-setting fires nearly every year. During the summer months, news reports are filled with stories of fires so large and intense as to be virtually unfightable. In spite of the costs and the odds, federal money continues to pour in as firefighters risk their lives to protect homes and businesses in the wildland-urban interface. Before we discuss how to get out of this situation, perhaps we should take a few moments to examine, in more detail, how we got into it.[8]

The landscapes of the American West have been shaped by fire. Lightning provides abundant ignitions each summer, and many Native American groups supplemented nature's inputs by purposely lighting fires. Early American settlers followed this practice, primarily as a means of removing fuels and reducing the likelihood of a high-intensity fire at an inconvenient time. Throughout the nineteenth century, smoke was as common in the forest as it was in the newly industrialized cities. The American public was interested in suppressing only the few fires that directly threatened lives or property.

Late in the nineteenth century, however, a German-educated forester named Gifford Pinchot burst onto the American political scene and started changing ideas about fire. Pinchot was a confidant of Teddy Roosevelt and the first chief of the federal Forest Service. He was instrumental in convincing Roosevelt to transfer control of the recently created forest reserves from the Interior Department's General Land Office to the Agriculture Department. When he accomplished this in 1897, these massive reserves were placed under his own control.

Pinchot's European education left him convinced that fire was an evil element and must be banished from the forest. Along with other German-trained foresters (the United States had no forestry school at the time), he pushed the idea of protecting the forest from the "light burning" practices of Native Americans and early settlers. Foresters denigrated the practice but made only a minor dent in the public's perception that fire was a perfectly reasonable and useful land management tool. In fact, President William Taft fired Pinchot in January 1910, in part because Pinchot appeared to be losing the political battle to protect the forest from flames. Though Pinchot lost the battle, paradoxically he won the ideological war only seven months later, when a series of seemingly unrelated events suddenly turned the political tables.

In August of 1910, the great American pragmatist philosopher William James published his final essay, "The Moral Equivalent of War," and then promptly died. The essay called on humans to invest their energies in battling nature instead of other humans. The day James died, August 26, 1910, smoke from massive fires on the other side of the continent drifted into his New Hampshire home and nearly obscured the sun.

The Great Idaho Fires of August 1910 were a defining event for fire policy and management, indeed for the policy and management of all natural resources in the United States. Often called the Big Blowup, the complex of fires consumed 3 million acres of valuable timber in northern Idaho and western Montana. The fires wiped entire towns off the map and killed seventy-eight firefighters, mostly in a thirty-six-hour period of gale-force winds that ended just four days before the death of William James. James's essay and the attention his death brought to it provided fuel for the sparks Pinchot and his colleagues had scattered, and the Great Idaho Fires propelled the idea of fire suppression across the American landscape. These fires were a defining event in the history of fire policy and management. Losses in subsequent fires and fire seasons are still compared to the timber, towns, and people lost in the summer of 1910. The battle cry of foresters and philosophers that year was simple and compelling: fires are evil, and they must be banished from the earth.

The federal Weeks Act, which had been stalled in Congress for years, passed in February 1911. This law drastically expanded the Forest Service and established cooperative federal-state programs in fire control. It marked the beginning of federal fire-suppression efforts and effectively brought an end to light burning practices across most of the country. The prompt suppression of wildland fires by government agencies became a national paradigm and a national policy. Subsequent events in the first half of the twentieth century reinforced the paradigm of fire suppression, and the rare voice of dissent was quickly drowned out. The 1935 meeting of the Society of American Foresters, for example, argued that fires were useful and perhaps even essential for the maintenance of some forested ecosystems. Influential foresters had concluded that the dogged pursuit of fire suppression, especially as it required the creation of new roads and trails, was reducing the cultural value of lands to a greater extent than the fires themselves. Apparently unimpressed, the Forest Service responded by implementing the "10 A.M. policy" that would drive fire management for four decades: the new national goal was to extinguish every new fire by 10 A.M. the day after it was reported. The nation's policy of aggressive fire suppression had begun in earnest. The Forest Service (under the U.S. Department of Agriculture) effectively became the national wildland fire-management agency; for the next several decades, it would not be an exaggeration to say that fire suppression was the primary goal of the Forest Service.[9]

THE ECOLOGICAL AND ECONOMIC COSTS OF FIRE SUPPRESSION

A disturbing subplot underlies the story of widespread federal fire suppression in the United States, in which scientific evidence is suppressed and the influence of scientists reigned in when their findings are not consistent with political agendas. This is exactly what happened in the Forest Service in the early 1930s, when its own research confirmed the positive effects of prescribed burning.

The Service assumed, perhaps correctly, that it would lose political influence and financial resources if it admitted that fires were beneficial in some systems. Some critics of federal fire management argue that the same pattern of suppression continues even today. Ultimately, of course, the Forest Service could not fool the forests, and the ill-advised policy caught up with them in a variety of ways.

What, then, has been the fallout from this program of complete suppression? From a strictly pragmatic perspective, it has failed rather dramatically in its goal to remove fire from wildland ecosystems. As we have seen, the number of acres burned each year is increasing, even though the number of fires has steadily declined since the early 1980s.

It is unlikely that suppression efforts have had much effect on high-severity fire regimes, in which many fires are simply uncontrollable, forests are normally dense, and fire-return intervals are very long. In the semiarid Southwest, however, where many forests evolved with frequent, low-severity fires, the picture looks very different. Since the mid-1970s, the area burned by wildland fires in Arizona and New Mexico has increased steadily and far more drastically than in the rest of the western states. While other factors, such as changes in land use and increased precipitation, likely contributed to this rise in fire activity, it now seems clear that fire suppression has played a major role.[10] In low-severity forests that developed with frequent surface fires, fire suppression has indeed caused an atypical accumulation of fuels. This accumulation has, in turn, led to an increase in very large, intense crown fires. Many large fires exhibit extreme behavior and spread quickly, so they can be very difficult to control.

Changes in fire regimes are not the only ecological consequences of fire suppression. As foresters of the 1930s astutely observed, aggressive fire suppression can affect ecosystems even more than the fires themselves do. When firefighters enter an ecosystem, they often begin by establishing fire lines, areas where vegetation and organic matter are cleared down to the mineral soil to keep fires from advancing, and where the firefighters establish a base for operations. New roads, helicopter landing sites, and campsites are often constructed. Whenever possible, fire crews will use preexisting roads and natural barriers, but lines may be constructed with hand tools; heavy equipment such as trenchers, plows, and bulldozers; or even explosives. These activities can compact the soil, severely erode it, and contaminate it, and heavily affected areas may require extensive rehabilitation afterward. Heavy vehicle traffic often spreads nonnative species. Suppression efforts often use water from nearby lakes, ponds, and reservoirs by means of pumps or helibuckets. The latter can remove up to two thousand gallons of water at a time, so the consequences for aquatic ecosystems can be severe.[11] Fire-suppression efforts sometimes take to the air, dumping millions of gallons of chemicals in the United States each year. These include fire-suppressant foams, which are usually detergent-based, and fire-retardant chemicals, which are usually nitrogen- or phosophorus-based.[12] Literature on the environmental effects of these chemicals is fairly limited, but they are known to be toxic to aquatic flora and fauna at higher concentrations.[13] Sizeable fish kills have been associated with the accidental application of fire-retardant chemicals over bodies of water.[14]

Even the use of fire to fight fire has the potential to damage ecosystems. In many cases, backfires lit to burn ahead of and help contain wildfires make up a substantial portion of the total area burned by a fire. The case of the 2002 Cerro Grande Fire in New

Mexico is a telling example. That fire, which burned into the town of Los Alamos and threatened the Los Alamos National Laboratory, was not exactly an escaped prescribed fire as is often reported. The fire that escaped and burned out of control was actually a backfire hastily lit in attempt to contain the original prescribed fire.[15] While we might argue that the negative ecological effects of such a fire are both inevitable (the area would have burned sooner or later under any management regime) and counterbalanced by the positive effects, it is still wise to recognize them.

The methods and results described here affect millions of acres of wildland ecosystems every year. In their militant efforts to contain fires and prevent all property damage, federal agencies are degrading the very lands they aim to protect. Fire, a natural disturbance with which most species have evolved, is replaced by the very unnatural disturbances of bulldozers, roads, helicopters, chemical retardants and suppressants, and thousands of firefighters. Where suppression's consequences are not greater than those of the fire, they certainly exacerbate them. For example, where soil erosion from a severe fire is worsened by damage from heavy equipment, the negative effects on soils, water, vegetation, and wildlife are multiplied.

The social, political, and economic effects of fire suppression policy are less visible but no less severe. Of course, the most obvious practical problem with fire suppression is that it is always ineffective in the long term and is often ineffective even in the short term. Much of the continuing emphasis on fire suppression today is based on the idea that putting out small fires is the only way to prevent them from becoming large fires and eventually threatening human communities. However, this practice permits the buildup of fuels that agency officials and politicians so often describe as having reached a "crisis" level. An area that is prevented by suppression efforts from burning will eventually be subject to fire again, and if fuels have continued to build up (and if the area's fires are generally driven by fuel availability), the next fire may well prove to be even bigger and more intense. Ultimately, a policy that spends $1 billion or more each year to worsen the problem it aims to solve is a failed policy. It is also a terrible drag on social, economic, and political systems. It wastes money that could be used more productively elsewhere, and it undermines the public's faith in policy makers and land managers.

This is particularly evident in light of the fact that suppression techniques are rarely effective, even in the short term, on large or intense fires. Most suppression efforts succeed only when a change in weather patterns brings increased humidity, lower wind speeds, or rain, or when a fire literally runs out of fuel.[16] A publication by the Western Fire Ecology Center quotes a Forest Service fire manager as saying, "Often we use resources because of the public and political pressure to do something, even though it has no effect on the fire and is an economic waste."[17] With the costs of suppression so incredibly high, as President George W. Bush's 2003 budget message noted, "in some western areas, the government pays more in suppressing fires than the fair market value of the structures threatened by those fires. It would literally be cheaper to let the fires burn and pay 100 percent of the rebuilding cost."[18] With the number of firefighter deaths and accidents growing every year, we ought to think twice before asking thousands of people to risk their lives for this increasingly dangerous and ultimately counterproductive endeavor.

To the very limited extent that fire suppression does temporarily protect human development from wildland fire, it indirectly subsidizes development in the wildland-urban interface. Much of the inherent cost of building in a fire-prone environment is borne by federal taxpayers rather than by developers or residents. By protecting the wildland-urban interface at no cost to the people who choose to build there, the federal government is effectively spending roughly $1 billion per year to encourage people to develop and live in fire-prone areas.

One of the oddest and most surprising traits of historical fire suppression policy is that, for much of the twentieth century, suppression activities were effectively given an unlimited budget. As a result of the 1908 Forest Fires Emergency Act, the Forest Service could use any available funds to fight "emergency" fires; it was not limited to funds designated for that purpose.[19] At first glance, this policy seems to make sense. After all, the stakes of fire suppression were seen as being so high that virtually no fire manager would want to stop fighting a fire when resources ran out. Furthermore, fire seasons are notoriously unpredictable: the Forest Service might fail to use a large budget one year and then spend it in the first month of the fire season the next. When it was assumed that most fires should be fought, it was, of course, practically impossible to prioritize suppression efforts and stay within a budget. In light of our new understanding of fire, however, we should closely examine the incentives and disincentives created by this system and its modern counterparts.

Congress repealed the "blank check" law in 1978 and tried to force the Forest Service to stick to its budget. For about ten years, the agency would borrow from other programs to pay for big fire seasons and then repay the difference when conditions were less severe. In the late 1980s, though, these other programs were so effectively drained that Congress gave the Forest Service a large supplemental appropriation. Ever since then, the Forest Service has been asked to operate within a fixed suppression budget, but it is allowed to continue borrowing from other programs and then draw money from an "emergency" contingency fund to restore these funds. It has drawn from this fund every year since 1993.[20]

When the Forest Service is permitted to borrow funds from other programs, it creates a perverse incentive to favor fire suppression over other fire and resource management activities, including fire use, fuel reduction, and even research and completely unrelated management programs. Although these funds are eventually restored, the current budget system assures that funds will always be available for even the most expensive and the least effective fire suppression efforts. Meanwhile, every other form of management takes a budgetary back seat.

Most people who live and work with wildland fire—residents of the wildland-urban interface, land and resource managers, policy makers, scientists, and the general public—are developing an increasingly nuanced understanding of it. There is broad consensus among the informed public and managers of natural resources that fire is necessary for the maintenance of many ecosystems, and that fires cannot always be extinguished or prevented. Still, efforts to translate scientific understanding of fire to the policy realm have met with great resistance. Decades have passed since the myth of 100 percent fire suppression was effectively debunked, yet fire policy and management continue to center on suppression efforts. For example, the Clinton administration's

National Fire Plan heavily emphasized the role fire suppression has played in increased fire severity, yet roughly half of the plan's budget was appropriated to suppression activities.[21] As one might expect, the language required to rationalize this decision is contradictory, stressing both the importance of aggressive suppression and the idea that "problem" fires are the result of these same activities.

A similar trend can be found in the George W. Bush administration's Healthy Forests Initiative. In a reversal of the general trend toward acknowledging fire's importance and complexity, the Healthy Forests Initiative paints the "fire crisis" with a dangerously broad brush. Although it claims to offer a better alternative to fire suppression, this paradigm actually reinforces many of the misunderstandings and protracts many of the ill effects of that older policy.

FOUR

Logging the Forests to Save Them

In the summer of 2002, conditions along the Colorado Front Range of the Rocky Mountains were ripe for a major fire. Unusually dry weather conditions had persisted in the area since 1998. In the winter of 2001–2002, a La Niña pattern in the eastern Pacific worsened the already low levels of precipitation and humidity. Dead fuels in particular were extremely dry, with some below 5 percent moisture content. After some brief, weak rains the first week of June 2002, the Front Range began to experience a weather phenomenon known as "summer blocking": the normal summer circulation patterns were interrupted by an atmospheric disruption to the west. This centered a high-pressure system over the Rockies, causing a rapid increase in temperatures and a further decrease in humidity.[1] This same phenomenon brought windy conditions, with winds gusting more than thirty miles per hour at times. On the afternoon of June 8, a single ignition occurred in the Taryall Mountains between Denver and Colorado Springs, and the Hayman Fire was born. Despite early and aggressive attempts at suppression, the fire could not be brought under control. It spread rapidly; by the following morning, it had already covered more than a thousand acres.

In the days that followed, the fire's behavior and spread were closely tied to the weather. The accumulated fuels, which often formed ladders from the ground up to the tree canopies, and which were fairly continuous downwind from the ignition site, likely played a major role as well, but weather appeared to drive the fire. Hot, dry, and windy weather continued for the first few days. For forty-six of the first fifty-two hours, relative humidity remained below 10 percent. On June 9, winds gusted to fifty-one miles per hour. The fire showed many typical signs of extreme fire behavior. It "spotted," spreading well ahead of its own front through flying embers, developed its own weather system in the form of high winds and pyrocumulus clouds, and spread across some sixty thousand acres that day in a single wind-driven run along the South Platte River. Crowning was widespread. The following day the weather calmed, and fire behavior moderated for about a week, although the fire did continue to spread somewhat. On July 17, low humidity and high winds returned, and fire intensity once again increased. A shift in wind direction further increased the area burned; on the 17th and 18th, the fire spread across an additional twenty-three thousand acres.[2]

The Hayman Fire's behavior was closely related to the kind of landscape it was burning, a midelevation mixed-conifer forest that included ponderosa pine, Douglas-fir, and blue spruce, among other species.[3] As we have seen, mixed-conifer forests typically have mixed-severity fire regimes, in which fuels and weather contribute to fire occurrence, severity, and behavior. These forests evolved with both low- and high-severity fires at varying return intervals. This mixed fire regime is reflected in the behavior of the Hayman Fire, which was at times driven primarily by fuels, at times by weather. Several previous fires and fuel-reduction projects had occurred in the burn area within the past several years, creating areas where fuel density and structure were altered. The Hayman Fire effectively blew through many, but not all, of the treated areas during periods of extreme weather. It ran virtually unchecked through areas that had been burned by wildfires in 1963 and by prescribed fires in 1990, 1992, 1995, and 1998. Areas where fuels had been mechanically reduced sometimes modified fire behavior by reducing flame height, but the fire remained intense on very dry, windy days. On calmer, cooler, and more humid days, however, recently treated areas appeared to reduce fire intensity and moderate behavior. With few exceptions, treated areas did not stop the fire from spreading in a given direction.[4]

In earlier chapters, we addressed the idea that our nation's history of fire suppression has led directly to increased wildland fire activity. We have shown that this has certainly been the case in some ecosystems previously characterized by frequent, low-severity fires. In these systems, fire suppression has led to an abnormal accumulation of woody material and a laddered fuel structure. These factors are indeed changing some fire regimes. In other systems, the previous one hundred years of fire suppression is unlikely to have had much of an effect, because fires have historically occurred in them relatively infrequently. In a forest where fires occur only every few centuries, a century of fire suppression is unlikely to have much of an effect on fuel accumulation and therefore fire behavior.

Unfortunately, policy makers under the last few administrations have generally failed to grasp this distinction. As a result, fire policy in the United States since the turn of the twenty-first century has seen small-diameter woody fuels (i.e., dense or "overstocked" forests) as the cause of nearly all damaging fires. This has led, predictably, to the impression that reducing hazardous fuels is a panacea for the problem of extreme fires. Under this fuels-centered paradigm, mechanical thinning and, to a lesser extent, prescribed burning are promoted in nearly all ecosystems. The popularity and overly broad application of this idea has led to many dangerously simplistic policy and management decisions.

The application of this fuels-centered paradigm in the policy realm is a case of massive overgeneralization in the transition from science to policy. Policy makers took the problems and solutions from the well-studied low-severity systems and created a set of policies for a wide variety of ecosystems and fire regimes across the United States. By applying ecosystem-specific research on too broad a scale, they effectively adopted a set of underlying assumptions that are not applicable in the vast majority of fire regimes. The ecological, political, social, and economic impacts of this oversimplification are only now becoming apparent, but it is safe to assume that, if this paradigm continues to be broadly applied, the negative results will rival those of the fire-suppression paradigm. Before we consider these results, it will be useful to examine the evolution of the paradigm and the policy decisions.

FIRE POLICY ENTERS THE TWENTY-FIRST CENTURY

In 2002, President George W. Bush and agency leaders presented their answer to wildland fire: the Healthy Forests Initiative, "an initiative for wildfire prevention and stronger communities." The initiative emphasized the extreme, "catastrophic" nature of recent wildland fires, stating that "enhanced measures are needed to reduce the risk of these catastrophic wildland fires."[5] The initiative explicitly advocated fuel reduction not only as a means of protecting communities from wildfire but also as a necessary tool for restoring the health of mismanaged forests. It started from the assumption that most public lands carry excessive and dangerous fuel loads, and then proposed that fire-management strategies should focus on mechanical fuel treatments and prescribed fire to reduce fuel loads as quickly as possible.

The Healthy Forests Initiative assumes that dense, fire-prone forests are inherently unhealthy and unnatural. Again, this assumption fits many ponderosa pine forests and other ecosystems with low-severity fire regimes quite well, but it overlooks the fact that many ecosystems do not currently carry excess fuel loads. Many western fire regimes are not driven primarily by fuel accumulations, and many have not been significantly affected by fire suppression policies. In these fire regimes, certainly, fuel-reduction projects are not a viable solution either for protecting communities from fire or for restoring ecosystem health, and "catastrophic" fires are neither unnatural nor avoidable. Contrary to popular perceptions, uncharacteristically dense, fire-prone forests do not blanket the western United States. As our earlier chapters illustrated, where dense, fire-prone forests do exist, fire suppression is not necessarily the primary cause. Dense forests may be a perfectly natural phenomenon (as in the case of coastal Douglas-fir forests and others with high-severity fire regimes), or they may be caused by a wide range of human activities.[6]

In an attempt to deal with the high level of variability among fire regimes, the federal land management agencies have developed a tool for determining the expected natural fire regime of a given site and the degree to which current conditions depart from that regime.[7] This would seem to be a promising tool, although the underlying idea of a "natural" fire regime, frozen in time and space, as a management goal is somewhat problematic. Several of the Healthy Forests Initiative policies refer to the classifications that result from this system and confine the application of these classifications to fuel reduction projects in the most affected forests (that is, those with the greatest departure from the natural fire regime).

Unfortunately, the fire regime classes and fire regime condition classes that form the core of this system are extremely broad. The natural fire regime of every ecosystem falls into one of five categories, based only on historical fire frequency; there is no consideration of fire intensity, seasonality, variation in frequency, or any of the other factors we have discussed here. The fire regime condition classes, used to indicate the relative departure from natural fire regimes, are even broader: Three classes have been defined to encompass every parcel of federal land in the country. An area may be described as experiencing a low (class 1), moderate (class 2), or high (class 3) departure from the average historical fire regime. The classes employed by this system are far too broad for site-specific forest management; in fact, the scientists who mapped federal lands into these classes specifically stated that they were too coarse to serve as a basis for management decisions.[8] Furthermore, this system tends to toss most sites into the same condition category (class 3). Resource-management agencies pay inordinate attention to these "worst-case" sites, which means that relatively little attention is paid to maintaining high-quality sites in condition classes 1 and 2. Focusing management efforts on these relatively intact systems would be considerably less expensive than prioritizing sites in condition class 3. It would also allow us to maintain large areas of healthy fire-prone ecosystems, rather than focusing our efforts on trying to rebuild damaged ecosystems at the risk of letting more areas deteriorate.

An important catchphrase emerged at about the same time that the Healthy Forests Initiative focused attention on excess fuel accumulation. The idea of an "analysis paralysis" became a recurring theme in the Healthy Forests dialogue and represented the political side of the fire debate. "Analysis paralysis" refers to the idea that the procedural requirements of environmental and land-management laws cause dangerous delays in the reduction of hazardous fuels. Reducing fuels in forests across the country (and especially on public lands in the West) was seen as such a pressing need that environmental laws were presented as a direct threat to public safety. In a 2002 USDA Forest Service report, the same concept was retermed a "process predicament," and the argument was expanded to include the claim that the public and environmental groups were further delaying much-needed projects through the excessive use of administrative appeals and litigation.[9]

These basic elements virtually dominated the discussion of fire ecology and management at the beginning of the twenty-first century: forests made unhealthy and unnaturally flammable by mismanagement; the desperate need to thin forests and save them from extreme wildfires; the hindrance of fire management efforts by red tape and burdensome regulations; and the inappropriate use of the appeals and litigation process by environmental groups bent on stopping sensible fuels management. The stage was set for a major policy shift, and the federal government adopted a new paradigm for dealing with fire.

A series of policy and regulation changes associated with the Healthy Forests Initiative have repeatedly stressed the same points. Primary among these is the Healthy Forests Restoration Act (HFRA). Somewhat ironically, the act was pushed through on the heels of large, home-destroying fires in California chaparral—fires that would be virtually impossible to prevent or mitigate through fuel-reduction projects. The irony seemed to escape agency officials. The HFRA aimed primarily to streamline the approval process for certain kinds of fuel-reduction projects and limit the potential for administrative appeals of these projects. Again, the argument was that environmental policies and appeals by misguided citizen groups were causing dangerous delays in fuel-reduction projects. The act also established grants for innovative uses of small-diameter wood, which is the primary by-product of most fuel-reduction projects. Thus, while the HFRA facilitated fuel reduction both through mechanical thinning and through fire use, the dialogue surrounding the passage of the act focused heavily on mechanical thinning. The administration argued that the needed projects could be funded at least in part by selling off the "excess" wood.

Similarly, two important rule changes under the umbrella of the Healthy Forests Restoration Act established alternative, less demanding processes through which some kinds of fuel-reduction projects could comply with the Endangered Species Act and the National Environmental Policy Act.[10] The former policy was seen as standing in the way of fuel reduction in areas where threatened or endangered species might be present. The latter forced agencies to conduct a fairly thorough analysis of the potential environmental effects of their thinning operations. Under rule changes, the regulatory burden for some fuel-reduction projects was substantially reduced; therefore, the potential for oversight and for public participation was likewise reduced. The stated goal of these changes was to resolve the "process predicament" for fire management activities that the federal agencies consider to be both crucial and relatively benign from an ecological perspective. The idea that small fuel-reduction projects are environmentally benign, at least when compared to large wildfires, would become a key theme in Healthy Forests policy. We discuss the environmental and social impacts of these changes later in the chapter. For now, it is worth noting that these two major policies form the bedrock of federal environmental regulations and are crucial in protecting both natural ecosystems and the American public from harm. Any attempt to bypass or undermine these regulations should be subject to very careful scrutiny. Indeed, thousands of concerned citizens wrote to comment on the draft regulations, many of them highly critical of the changes.

As we have seen, the Healthy Forests Initiative and its associated policies and regulations depend on a limited and narrow understanding of wildland fire, one which, applied very generally, suggests that intense or large "problem" fires can be prevented by actively managing fuels. Because this administration's initiative has failed to incorporate a more site-specific understanding of fire, the policies enacted under it dangerously underestimate the complexity and importance of wildland fire. In this sense, these policies parallel the historical policies of aggressive federal firefighting. The result is a national system of fire policy that is dysfunctional and even harmful in many local areas.

The idea that fuel-reduction treatments are environmentally benign when compared to wildland fire in its uncontrolled state forms the logical basis of many of the policy changes related to the Healthy Forests Initiative. This idea provided much of the rationale for expediting the approval of such treatments, and it is also a common theme in statements by the president, cabinet members, and agency officials. Certainly when these individuals appear at the site of a recent wildfire and stand in front of a blackened landscape, it is easy to believe that any amount of mechanical thinning would be preferable to a large fire. Unfortunately, the nature and impacts of these projects vary greatly, and in many cases the assumption about lesser harm is questionable or even patently false. Like early fire-suppression policies, the Healthy Forests Initiative and related policy and regulation changes have the potential to seriously harm western ecosystems. Moreover, the limits they set on appeals and public participation, we argue, also undermine the political and social structures that support democratic systems.

HEALTHY FORESTS: RESULTS AND CONSEQUENCES

As we have seen, fire plays an important and beneficial role in most western ecosystems, so the comparison between the effects of fuel-reduction treatments and of wildland fire is somewhat flawed from the beginning. The Healthy Forests policies promote both mechanical treatments (thinning) and prescribed fire as tools for reducing hazardous fuels, although both opposition to and support for the policies have focused primarily on mechanical thinning projects. We focus here primarily on the effects of the more contested activities.

In describing the ecological effects of mechanical thinning projects, we are forced to consider how these projects differ from commercial logging operations. Commercial logging is, of course, associated with a plethora of negative environmental impacts and cannot be described as ecologically benign on virtually any scale. While this question is understandably contentious in forestry circles, the authoritative dictionary of forestry terms and at least two popular silviculture textbooks agree that mechanical thinning effectively differs from logging only in intention.[11] In other words, mechanical thinning is defined as a logging operation whose primary intent, when applied to fire management, is to change fire occurrence or behavior.

As an example, the strategy proposed by Northern Arizona University's influential Ecological Restoration Institute (which provided much of the research used to support the Healthy Forests policies) to restore southwestern ponderosa pine ecosystems includes very heavy "thinning from below." This strategy removes a high proportion of small, young trees, and it produces a stand structure similar to that produced by a normal seed-tree timber harvest (see figure 8).[12] It is also what most people tend to picture when they hear of fuel-reduction or mechanical thinning projects. But some other mechanical thinning projects involve practices that closely resemble logging techniques. Each of these timber-removal techniques, whether intended for profit or fuel reduction, represents a point along a continuum of tree removal, with clear-cutting representing complete removal of trees, seed-tree harvest or heavy thinning representing nearly complete removal, and so on through the removal of no trees at all.

In policy documents, projects to be included under Healthy Forests laws and regulations are almost exclusively defined as those where the "primary purpose" is hazardous fuels reduction.[13] Agencies and policy makers argue that, by their very definition,

Figure 8.   Site of a seed-tree cut in a Washington State ponderosa pine forest. Photo by Guy McPherson.

mechanical treatments target smaller trees while leaving larger trees and (sometimes) snags. This interpretation is consistent with public perceptions of thinning and timber harvesting but not necessarily with contemporary definitions or uses of those terms. In fact, the policy documents associated with the Healthy Forests Initiative address the perceived distinction between thinning and timber harvesting in only the vaguest terms. For example, projects covered by the Healthy Forests Restoration Act are required to focus "*largely* on small diameter trees, thinning, strategic fuel breaks, and prescribed fire to modify fire behavior" and to maximize "the retention of large trees . . . *to the extent that the trees promote fire-resilient stands*" (italics added).[14] In other words, mechanical thinning projects are considered ecologically benign, and qualify for an expedited review process, if they mainly remove smaller diameter trees and if they reduce the overall flam-mability of the forest. Aside from the obvious question of whether forests that have always been shaped by fire can or should be made "fire-resilient," the unwillingness of policy makers to clearly define the permitted activities is worrying. It is not at all clear from the HFRA's wording that thinning would be com-pletely limited to small trees, or that the activities and their effects are likely to differ substantially from those of timber-harvesting operations.

The kinds of fuels removed by a given project largely deter-mine the project's likely effects on fire occurrence and behavior. Mechanical treatments that remove larger trees, and which are commonly and appropriately referred to as "thinning," can greatly increase fire frequency and intensity.[15] The distinction between the removal of small versus large fuels should be addressed clearly in fire management policy. This is the only way to ensure that fuel management matches public expectations.

The negative ecological impacts of commercial logging are well documented, and they combine to seriously undermine crit-ical ecosystem functions. Among the most often-cited impacts are the introduction of nonnative species, destruction and dis-ruption of wildlife habitat, stream sedimentation, and substantial damage to soils, especially where new road construction is re-quired.[16] Many of the policy changes associated with the Healthy Forests Initiative assume that fuel-reduction efforts will not cre-ate these negative effects, but it is not at all clear how such effects can be prevented. Prescribed fires can produce similarly unde-sirable ecological consequences, especially when they are used uncritically and without an informed, site-specific analysis of likely effects. For example, a prescribed fire conducted outside of the normal fire season may interrupt plant or animal reproductive cycles, cause greater soil erosion, or even increase the risk of wildfire in cases where an incomplete burn leaves behind a large amount of dead vegetation.

The effects of fuel-reduction projects on soils are especially worrisome, in part because soil that erodes from a site is very difficult and costly to replace. Mechanical treatments in particular accelerate soil erosion, although stand-replacing fires also often lead to massive soil movement and sedimentation. In each case, soil is removed from a productive site and deposited in waterways, diminishing the productive capacity of the original site and eliminating habitat for aquatic species. Soil development in forests and grasslands occurs so slowly that it is usually described in terms of centuries per inch of soil. As a result, erosion-induced loss of a site's productive capacity is typically viewed as irreversible. Forests and grasslands simply cannot be productive for any species (including humans) when they are stripped of their soil.

Mechanical treatments also typically reduce the productivity of soils that remain on-site. Heavy machines compact soil. The resulting decline in productivity has been recognized as a problem with mechanical logging since at least 1928, and the dangers of compaction have been reiterated in dozens of forest ecology textbooks since then. A voluminous literature on this topic is summarized by the renowned forest ecologist David Perry in his 1994 text, *Forest Ecosystems:* "Heavy machinery compacts soils, which reduces porespace and therefore water infiltration, aeration, and the ability of plants to root effectively. As little as one pass of heavy equipment over a piece of ground can compact soils sufficiently to reduce tree growth, and effects can persist for decades."[17] The resulting impediment to water infiltration causes rainfall to run off more quickly than from uncompacted soils, further accelerating soil erosion. Soil compaction is often perceived to be a minor, unavoidable cost of mechanical treatments, but its effects are neither inconsequential nor short-lived.

The effects of fuel-reduction projects on aquatic ecosystems are similarly dramatic. Thinning treatments and conventional harvesting strategies change water chemistry and contaminate aquatic ecosystems by eroding soil. Increases in mercury levels in freshwater fish have been attributed to timber-harvesting practices. A series of case studies in Canada led researchers to conclude that "emulating" fires through mechanical treatments is "not appropriate . . . to protect and preserve aquatic ecosystem integrity, since the nature and intensity of some impacts . . . differ between wildfire and logging." Researchers further concluded that buffer strips, a common device for mitigating the effect of mechanical treatments, failed to protect aquatic ecosystems.[18]

In isolated areas with intact wilderness, the most visible result of mechanical-thinning projects is the fragmentation of ecosystems. Permanent or temporary roads are often built in previously undisturbed areas, and brush, small trees, and other coarse woody debris are generally removed, threatening the many species that depend on them for habitat. When left in place, coarse woody debris and other "hazardous" fuels create a mosaic of highly complex and productive microenvironments that play an important role in maintaining biological diversity. They are not, as many policy makers seem to believe, excess forest material that must be cleaned out to create a healthy ecosystem. For example, many species of birds and small mammals use snags preferentially or even exclusively for nesting, food storage, hunting, courtship, and other purposes. Many charismatic large animals, such as black bears, also make good use of dead trees.[19] Even the dense stands of small trees so often described as the antithesis of healthy forests provide habitat for some rare and beautiful species. For example, the Kirtland's warbler, a federally listed endangered species so well known yet so rare that it has a National Wildlife Refuge named specifically for it, nests exclusively in large, very dense stands of small jack pine *(Pinus banksiana)* trees.[20]

Endangered and threatened species present a particularly complex problem in these fragmented ecosystems. Before European settlement of North America, metapopulations served as sources of species recruitment into burned areas. That is, when a disturbance such as fire (whether natural or human-caused) demolished a local population of a species, the population generally was replaced by individuals from other nearby populations. However, fragmentation creates many obstacles to such filling in of extirpated populations from off-site. When wildlands become fragmented, there may be no nearby populations available to recolonize an area. The large-scale cycling of populations is interrupted, and isolated populations become highly vulnerable to even small-scale disruptions. As a result, any attempt to reduce fuels, whether through thinning or fire, may threaten the short-term survival of a species whose numbers have already been seriously diminished by other activities. Of course, the potential risks to rare species associated with fuel reduction must be balanced with risks to these species associated with no action. In many cases there may be no good option available for protecting vulnerable species. Instead, managers must carefully weigh the risks associated with different management plans.

Like fire suppression, attempts to control fuels on a large scale and create "fire-resilient" stands (to whatever extent that might be feasible) encourage further development of the wildland-urban interface and make continued fragmentation appear virtually risk-free. Thus, the use of public funds to control wildland fuels creates a perverse incentive for humans to increasingly inhabit areas that, realistically, will always be fire-prone to some extent. Efforts to create fire-resilient forests therefore actually create a positive-feedback cycle of ever-increasing ecosystem fragmentation.

Finally, and perhaps most importantly, we must recognize that creating "fire-resistant stands" is not the same as restoring ecosystems. The reckless or ill-considered use of either mechanical thinning or prescribed fire to control fire often creates an inadvertent threat to native biological diversity. Unfortunately, the irreversible fragmentation of ecosystems has virtually eliminated the possibility of restoring historic fire regimes in many areas.

On the whole, the negative ecological impacts of mechanical treatments can easily exceed those of wildfires, and they may persist long after any economic or social benefits are fully realized. First and foremost, we should remember that native species throughout most of North America evolved in the presence of recurrent fires, and they have developed strategies to avoid, escape, or tolerate periodic fires. Whereas evolution has ensured that native species are able to cope with, and in many cases benefit from, recurrent fires, the industrial age has been far too short to allow native species to develop adaptations to mechanical treatments. As much as we might like to replace unpredictable wildland fires with neatly trimmed fire-proof forests, nature will not be fooled.

Just as they threaten the well-being of wildland ecosystems, the policies developed thus far under the Healthy Forests Initiative also threaten to subvert key democratic processes. They limit citizens' involvement in public land management and discourage the kinds of active citizenship that are crucial to democratic decision-making. The Healthy Forests policies developed under the George W. Bush administration are built around the assumption that appeals, citizen litigation, and even regulatory policies in general are incompatible with effective fire management.

Manifestations of these ideas abound in the public dialogue, especially in speeches by agency leaders. Mechanical thinning is seen as a virtual panacea in restoring historical fire regimes, and federal environmental policies that can delay or prevent thinning activities in sensitive areas are routinely described as "unnecessary regulatory obstacles," "layers of unnecessary red tape and procedural delay," or "burdensome regulations."[21] The 2003 "Process Predicament" report contends that "unfortunately, the Forest Service operates within a statutory, regulatory, and administrative framework that has kept the agency from effectively addressing rapid declines in forest health."[22] At President George W. Bush's signing of the Healthy Forests Restoration Act, he commented, "The bill expedites the environmental review process so we can move forward more quickly on projects that restore forests to good health. We don't want our intentions bogged down by regulations."[23] On several occasions political proponents of this idea have alleged that citizen groups and environmental laws were responsible for catastrophic fires and the loss of property, and even life, associated with them. Following the 2002 Rodeo-Chediski Fire, Arizona governor Jane Hull and Senator Jon Kyl directly implicated environmental groups, with Kyl stating that environmentalists "would rather the forests burn than to see sensible forest management."[24] Likewise, Representative Scott McInnis (co-sponsor of the Healthy Forests Restoration Act) claimed that the deaths of four firefighters in the 2001 Thirty Mile Fire were the direct result of the Endangered Species Act; this was later flatly refuted in the Forest Service's accident report.[25] Nonetheless, proponents of the "process predicament" idea argue that the current requirements of environmental and land management policies are far too complex and demanding, and that a desire to avoid appeal or litigation often causes land managers to produce unnecessarily thorough analyses.[26] Overall, there is an assumption that regulatory processes are too constraining and too time-consuming to be applied to fire management decisions.

These ideas, which virtually dominate fire policy at the beginning of the twenty-first century, appear to be based on a relatively few isolated cases. The regulations concerning implementation of policies like the National Environmental Policy Act and the Endangered Species Act clearly address timeliness and even allow agencies to set time limits for each part of the process based on the potential for harm, the magnitude of the action, the degree of controversy, and other criteria.[27] A report from the nonpartisan General Accounting Office suggests that litigation has played a relatively minor role in delaying hazardous fuels projects. Of 762 fuels-related decisions in fiscal years 2001 and 2002, only 180 were appealed at all, and only 23 were litigated.[28] The same report also found that 79 percent of appeals were handled within the mandated ninety-day period. A separate report found that 30 percent of the delays on fuels projects at sites visited by GAO researchers were caused by the reallocation of project funds to fire suppression efforts. As we noted earlier, weather accounted for an additional 40 percent of delays.[29]

The accuracy of these reports has been questioned, and the lack of transparency in Forest Service records may make a clear evaluation of the "process predicament" idea all but impossible. Ultimately, though, we should not be surprised if public participation and accountability do slow things down. Democracy has never been a simple or expeditious process, but it is democracy's very deliberative nature that we have historically found so valuable. Surely we do not want to throw aside the public's right to participate, to contend, to appeal poor decisions, or to litigate over perceived harm just because we would like the decision-making process to move more quickly.

How can we evaluate the impact that Healthy Forests policies have had on democratic processes? As policy researchers Hanna Cortner and Margaret Moote note, natural resource management in the United States has historically been characterized by a tension between expert and participatory management. At the same time, political and economic interests have generally dominated management processes under both regimes.[30] In other words, whether scientists or the public carry the theoretical mandate to make management decisions, the interests of political parties and corporations tend to dictate much of what actually happens on the ground. Environmental regulatory policies were developed primarily as a response to this domination of interests. People saw the havoc that competing interests were wreaking on the national landscape, and many felt the federal government should step in and at least moderate the level of damage. Many of the policies circumvented by new Healthy Forests policies were originally intended to make public land managers more responsive to citizens (and, presumably, less responsive to extractive industries). The National Environmental Policy Act (NEPA), for example, requires major decisions made by federal agencies to take environmental consequences into account, but the truly significant part of the policy is that it requires agencies to put this decision-making process into writing and make it open to public and judicial review at several levels. Likewise, the Appeals Reform Act (ARA) requires that a "notice and comment" process be established for proposed actions on national forests, and it permits any affected citizen to appeal decisions made by the USDA Forest Service.

Policies like NEPA and ARA have represented a substantial shift from expert to participatory management, because they explicitly allow citizens to participate in the management of public lands. The administrative appeals process set up in these and other policies provides a process through which the public can challenge bureaucratic decisions they see as harmful without resorting to costly, time-consuming litigation. While such structures are certainly not sufficient to ensure citizen participation, they do provide an important forum for citizens who see their interests threatened by agency decisions.

The Healthy Forests Restoration Act undermines participatory democracy in public land management by gutting the standard appeals process on administrative decisions. Projects covered by HFRA are explicitly exempted from the normal appeals process; the environmental assessments and environmental impact statements required under NEPA are not subject to the standard process for public review established through the ARA.[31] Not just anyone can challenge decisions through the new appeals process. The Secretary of Agriculture developed a special, much more demanding process for individuals or groups who want to force the administrative review of an HFRA fuel-reduction project. Citizens and citizen groups can participate in this review process only if they have submitted comments on the original draft analysis. They are not allowed to appeal after a decision is made, as would normally be permitted with non-HFRA projects. Instead, any appeal must be made within a thirty-day permitted filing period. Civil litigation generally can be pursued only when all administrative review and appeal procedures have been exhausted, and any court-ordered delays in litigated projects are limited to sixty days. Thus, the Healthy Forests Restoration Act creates a number of roadblocks to citizens' involvement in managing their public lands.

It is undisputed that these restrictions are intended to reduce citizen involvement in public lands management; the restrictions' proponents in the executive and legislative branches merely offer the rationale that frivolous appeals and litigation are hampering effective management. Ironically, the measures are often described in Healthy Forests documents as is if they strengthen public participation. For example, the White House Web site's description of the Healthy Forests Restoration Act states that it will "improve the public involvement in the review process by providing opportunities for earlier participation, thus accomplishing projects in a more timely fashion."[32] Nonetheless, the bulk of political rhetoric surrounding the initiative depicts citizen participation and citizen groups as frivolous and excessive. We argue that when federal officials malign democratic participation, this is in itself harmful to a democratic society.

As two noted forest scientists point out, current fire policies address only "procedural matters and do not address substantive issues such as where, how, and why fuel projects are to be conducted. The assumption appears to be that if we free resource managers from procedural constraints, they will make the appropriate decisions about where, how, and why."[33] Natural resource managers, fire managers, and the like are generally very knowledgeable, intensely familiar with the ecosystems they manage, and highly committed to responsible stewardship, but they should have a clear, cohesive fire policy that helps guide that stewardship, rather than one that merely frees them from the requirements of regulatory policies. In leaving agency personnel without this guidance, we force them to find their own way to balance the conflicting political, economic, and social pressures and create a viable set of goals and priorities for fire management.

Many analysts see the current policy environment as part of a much broader trend, the beginnings of a swing back to management by experts and technicians and away from public involvement. Executive orders signed by Ronald Reagan and Bill Clinton are often cited as initiating the recent trend against public participation laws. For example, Clinton's Executive Order 12866, which began "a program to reform and make more efficient the regulatory process," appears to have established much of the antiregulatory language invoked in Healthy Forests policies, and may also be seen as directly undermining the citizen-friendly Administrative Procedures Act, which provides for public involvement in agency rule-making.[34]

In 2003, while the Healthy Forests Initiative was the major focus of attention for many environmentalists, the trend against citizen participation continued as the Forest Service substantially revised its interpretation of the Appeals Reform Act and wrote new rules and processes to guide its implementation of the act.[35] These regulations apply to all appeals against Forest Service projects, not only to forest-thinning projects. Several changes were made that seriously undermined the ability of citizens to appeal Forest Service projects that might adversely affect them.[36] The new regulations made projects that had been excluded from analysis under the National Environmental Policy Act also exempt from the notice and comment and appeals procedures required by the Appeals Reform Act.[37] This includes the fuel-reduction projects outlined under the Healthy Forests regulation changes, as well as relatively small logging projects.[38] No environmental analysis or environmental impact statement is required of such projects. No legal notice or opportunity to comment is required. Concerned citizens cannot contest them. The regulations were also changed to allow many "emergency" projects to continue even while they are under appeal. The previous definition of these "emergency" activities was expanded to include cases in which there would be "substantial loss of economic value to the Federal Government if implementation of the decision were delayed"; it is probable that this change was made to include salvage logging operations.[39] The ultimate outcome of all of these changes is that fewer and fewer projects are subject to public scrutiny and appeal.

Because the provisions of the Healthy Forests policies shift so much discretion from citizens and the judicial and legislative branches to the executive branch (i.e., to federal agencies), opposition to them has come from a surprisingly disparate group of organizations. The groups opposing an early version of the Healthy Forest Restoration Act (passed by the Judiciary Committee of the House of Representatives) included not only environmental groups such as the Natural Resources Defense Council and the Wilderness Society but also a number of groups concerned with judicial independence and civil rights laws, including the ADA (Americans with Disabilities Act) Watch/National Coalition for Disability Rights, the National Association for the Advancement of Colored People, the National Organization for Women, Planned Parenthood Federation of America, and the Mexican American Legal

Defense and Educational Fund.[40] When the bill passed through the committee, the dissenting view was highly critical: "Proper administrative and judicial review of executive decisions and regulations are among the cornerstones of our system of government, which counts checks and balances as a basic tenet. This legislation attempts to eviserate *[sic]* these checks and balances to give cabinet and Federal agency officials virtually unchecked decisionmaking authority, seeks to subject plaintiffs and courts to rigid deadlines, and endeavors to place every Federal lawsuit except those pertaining to this legislation on the back burner."[41]

These threats to NEPA and the other policies and laws that provide a forum for citizen participation have prompted sharp criticism from legal experts. Sharon Buccino, an attorney for the Natural Resources Defense Council, describes NEPA as being "under assault" from critics in the executive and legislative branches and points out that the recent proposals "seek to circumvent the NEPA process, rather than improve it."[42] In an article in the New York University's respected law journal, she outlines the tactics being used to undermine opportunities for environmental review and public participation, and concludes, "As NEPA prescribes, government officials should be striving not simply to speed decisions up, but to make better decisions."[43]

The Healthy Forests policies also discourage active citizenship in less direct ways. Policy researchers Anne Schneider and Helen Ingram describe a situation in which public policy design has become increasingly politicized (they call it "hyperpoliticization"). They argue that many policies construct mutually oppositional groups in which the interests of different groups of citizens are pitted against one another. In doing so, they argue, "policies deceive, confuse, and in other ways discourage active citizenship, minimize the possibility of self-corrections, and perpetuate or exacerbate the very tendencies that produced dysfunctional public policies in the first place."[44]

The Healthy Forests Initiative and related policies have certainly created oppositional target groups, or have at least reinforced the divisions between preexisting groups. The divisive result of these policies is that environmentalists are pitted against business interests and even the "average citizen"—all to the ultimate end of increasing decision-making power in the executive branch. As noted earlier, the underlying assumption that environmentalists tie up the courts with frivolous lawsuits is pervasive in Healthy Forests documents. The blame placed on environmental groups for destructive fires is perhaps the most extreme example of this. Despite substantial evidence that the Forest Service's ongoing policy of aggressive fire suppression at any cost causes more delays in fuel-reduction projects than the actions of environmental groups (by usurping funding originally intended for nonsuppression activities), officials insist that "the agency often works diligently and collaboratively to design a project acceptable to constituents, only to have implementation stalled by a very small minority relying on esoteric legal arguments."[45] While announcing his Healthy Forests Initiative in Oregon in 2002, President Bush stated:

> And we have a problem with the regulatory body there in Washington. I mean, there's so many regulations, and so much red tape, that it takes a little bit of effort to ball up the efforts to make the forests healthy. And plus, there's just too many lawsuits, just endless litigation. . . . We want to make sure our citizens have the right to the courthouse. People ought to have a right to express themselves, no question

about it. But there's a fine balance between people expressing their selves and their opinions and using litigation to keep the United States of American from enacting common sense forest policy.[46]

Policy makers offer little evidence to support these assertions; as described by one policy analyst, the campaign to demonize environmental groups and limit appeals has been built on "the use of rhetoric, synecdoches, and repetition of unconfirmed data."[47] The extent to which these policies serve to deceive and confuse citizens remains to be seen, but it seems clear that restraints on citizen appeals and litigation are actively reducing the likelihood of policy self-correction. Bad decisions will occasionally occur in any bureaucratic program; it is crucial to the health of the democracy that such decisions be transparent and contestable.

In light of the negative ecological and political effects of Healthy Forests policies and practices, they must be recognized as a massively inefficient and inappropriate use of taxpayer dollars. These policies and practices promote overly general management of very complex and unpredictable ecosystems, they encourage continued development in fire-prone areas, and they undermine citizen participation in democratic decision-making.

Today there is substantial evidence that homes and communities can be well protected by managing fuels directly in the wildland-urban interface. If this is true, public lands management can give up the idea of creating fireproof forests and focus instead on conserving wildland ecosystems. In the next chapter, we provide evidence for this proposition and for the idea that fire itself is the most efficient and effective tool for managing most fire-prone ecosystems. Today, management practices that use fire are severely underfunded as a regular component of fire management, even as the Healthy Forests Restoration Act appropriates more than $3 billion over the fiscal years 2004 to 2008 to projects that are exempt from normal regulations and nearly impossible for citizens to appeal.[48]

FIVE

Tools for Living with Fire

In 2005 the fire suppression bill for the five federal land-management agencies—the USDA Forest Service, the National Park Service, the U.S. Fish and Wildlife Service, the Bureau of Land Management, and the Bureau of Indian Affairs—totaled $875 million. Federal and state agencies conducted prescribed burns on 2,310,346 acres and allowed several hundred lightning-ignited wildland fires to burn another 489,186 acres. Wildland fires burned a total of 8,686,753 acres across the nation that year. The obvious goal of this massive effort was to prevent large, destructive wildfires, the kinds of fires that eat up timber, blacken scenic vistas, and burn into populated areas. Efforts to prevent big fires focus on suppressing them while they are still small or remote, and on controlling fire behavior and spread by reducing available fuels. As we have seen, it is in no way possible to prevent all large, destructive fires. More important, our futile attempts to prevent such fires are both costly and ecologically and socially harmful, even though they are often promoted under the guise of building healthy ecosystems and healthy communities.

In order for our government to move toward more practical and progressive fire management, we suggest an explicit restatement of goals. Currently, the first stated priority of federal land-management agencies is to protect life and property, which is indeed laudable. Unfortunately, preventing and suppressing fires on 673 million acres of federally owned lands (nearly 30 percent of the United States) is neither an effective nor an efficient way of reaching that goal.[1] It just does not work. Managing fuels and fighting fires far from developed areas does not protect residents who live at the wildland-urban interface, and these activities put the lives of thousands of firefighters at great risk every year.[2] The good news is that we know how to better protect lives and private property, and we have the tools to do so.

Before we explain how to best achieve this goal, however, it will be useful to examine other goals and priorities for managing fire on western lands. Public lands, in particular, should be managed for the public good and not exclusively to enhance the safety and property value of nearby residents. The public uses wildlands for recreation, for gathering food, firewood, and Christmas trees, and for private logging and ranching operations. We depend on federal lands to provide nearby communities with clean water and air and to serve as a reservoir for biological diversity. The scenic vistas and recreational opportunities these lands provide substantially increase the values of adjoining property. Perhaps most important, many Americans take pride and comfort in the knowledge that we have protected some of our wildlands from development and will continue to protect them for future generations to enjoy. As a nation, we take great pride in our wildlands and make good use of them. Recreationists spend more than 1.2 billion days visiting federal lands each year.[3] Are we prepared to sacrifice the ecological richness of these lands to a misguided attempt to protect them from fire? Are we so determined to make our forests safe for a few people to live in that we are prepared to destroy them—and break the bank—in the process?

It is time to define a new set of goals for fire management. These goals should encompass the variety of values and services provided by wildlands. For moral, social, and political reasons, the first goal of fire management must remain the protection of human life. This means we cannot place the highest priority on property protection, because it so clearly comes into conflict with the protection of firefighters' lives. The current policy does not intend to sacrifice firefighters' lives for the protection of property, and fire managers make this clear at every available opportunity. Nonetheless, asking federal firefighters to place a high priority on protecting private property leads to unnecessary firefighter deaths, especially when managers choose for political reasons to fight fires they know they cannot even begin to control. As we have seen, there are dire ecological and social consequences associated with fighting wildland fires, and it is neither reasonable nor acceptable to sacrifice our publicly owned wildlands and political and economic well-being to protect private property. By putting the protection of private property ahead of other management concerns, we effectively manage our vast tracts of public wilderness for the benefit of a relatively few residents of the wildland-urban interface.

The second goal of fire management on public lands should be the maintenance and restoration of native biological diversity and the processes that sustain it. There is no shortage of practical and moral arguments in favor of maintaining and restoring biological diversity. Perhaps first and foremost, we know that native biological diversity is important because present and future generations of humans depend on a rich diversity of life for survival and civilization. As architects of the future of the planet, we have a responsibility to retain as much biological diversity as possible. The substantial economic cost of maintaining high levels of biological diversity pales in comparison to the costs of failing to do so. Restoring the processes that fostered biodiversity in the past is a huge step in the right direction. Inevitably, maintaining and restoring biodiversity in many western ecosystems will require the reintroduction of regular fire. For species that evolved with fire, and with specific kinds of fires and fire regimes, there can be no substitute.

The third and final goal of fire management should be to protect private property from fire *only* to the extent that it is possible to do so without compromising the first two goals. As we have shown in previous chapters, specifically prioritizing human life and biological diversity over property protection would be unnecessary were it not for the common misconceptions that fire suppression and mechanical thinning are the best ways to protect private property. We have seen how these practices both undermine existing systems and can actually increase fire risk in the wildland-urban interface. In this chapter we discuss how property in the wildland-urban interface can be protected most effectively, even while prioritizing the protection of human life and biological diversity of native species.

As we saw in chapter 1, we have three major tools at our disposal for managing fire. We can suppress it directly (to the best of our abilities), we can manage it indirectly by physically removing fuels, or we can use fire to our advantage, either by setting prescribed fires or letting already-ignited fires burn.[4]

Fire suppression tends to increase available fuels over time. It is a poor solution in the long term because it accelerates and intensifies some fire regimes and is harmful to most ecosystems. The way fire suppression has been used for most of the last century, as the preferred response to every unplanned fire, is counterproductive to our goals. It is improper to use public funds to suppress a fire on remote public lands purely for the protection of private logging or ranching interests or for the preventative defense of private real estate that may lie hundreds of miles away from the fire. This is especially true in ecosystems with a low-severity fire regime, where suppression is known to progressively increase the danger of a large, intense fire.

Nonetheless, there are times and places where a short-term solution to fire is preferable. Where wildland fire directly threatens human life, suppression is both justified and morally necessary. Fighting fire in the wildland-urban interface makes sense. Here, suppression efforts are most likely to be effective in preserving life and property, and they are least likely to damage intact ecosystems. However, fire management must advance (and slowly is advancing) beyond the automatic and aggressive suppression of any and all wildfires.

Mechanical fuel reduction, on the other hand, attempts to prevent fires that are large or intense, or that may threaten developed areas. Fire managers do this by altering the abundance, cover, or structure of fuels.[5] The most common mechanical treatment is thinning, which removes certain trees to improve the "quality" of the remaining stand of trees. Usually, thinning projects remove small-diameter trees to favor a lower density of large trees over a higher density of small ones. For example, many prescriptions remove many or all trees less than nine inches in diameter. However, thinning is also used to open the forest crown by removing larger trees ("high" or "crown" thinning), or managers may remove trees without regard for size, quality, or species to achieve a specified level of spacing or clumping.[6] Other methods of fuel reduction include the removal of understory shrubs and branches to prevent the accumulation of ladder fuels or the removal of needle litter from around the bases of trees or structures to prevent their ignition, although these practices are less common.[7]

Mechanical fuel reduction projects play a major role in the debate over fire management on public lands. Yet they are extremely costly, poorly matched to the evolutionary history of western ecosystems, and often ineffective. The indiscriminate removal of trees in a "one-size-fits-all" approach is a prescription for disaster. Even when only very small trees are removed, fuel-reduction projects should not be conducted without regard to the many important details. Managers should take into account the underlying causes of high stem densities, the fact that many dense forests are in no way overgrown or overstocked, and the importance of weather and climate phenomena in the development of extreme fire hazard. They should also cultivate an explicit appreciation for fundamental conflicts in the values and interests of different stakeholders.[8]

The small size of most of the trees removed by fuel-reduction projects makes them practically worthless from an economic standpoint. As noted earlier, the federal government has been promoting research into the uses of small-diameter fuels, but even if some uses are found, the proceeds of projects are unlikely to match the costs. Removing small-diameter trees from overgrown forests can cost more than five hundred dollars per acre.[9] At the national level in the United States, the agencies' definition of fire regime condition class suggests that the Forest Service should remove small trees from 50 million acres immediately, and then start working on another 80 million acres.[10] At that point, the Forest Service could breathe a bit more easily and start a program of maintenance that uses low-intensity prescribed fire. Other federal land-management agencies face the same problem at similar scales, which suggests that at least $100 billion would be needed to mechanically reduce fire hazard in the most fire-prone western forests. Admittedly, the fire regime condition class system probably overestimates the area in which forests are actually unnaturally dense, but this system remains the primary decision-making tool for federal-level policies. Certainly even the most optimistic land managers do not expect government agencies to commit funds at this level, and federal funding for land management and environmental protection continues to decline as funds are increasingly committed to armed conflicts and internal security. This lack of funding is exacerbated by the continued recruitment of small trees in western forests (trees will, of course, not stop reproducing, nor would we want them to), as well as by the regular annual reappropriation of management funds for emergency fire suppression. Mechanical fuel reduction is a fool's errand in many of the West's public lands, especially in light of the many other causes of accelerating fire regimes (such as climate and land use changes). Because of the vast areas involved and the rapid rate of regrowth in many ecosystems, it is difficult to imagine that mechanical treatments will prove to be an efficient tool on most lands.

The effectiveness of mechanical treatments is poorly documented for many ecosystems. Thinning cannot prevent crown fires completely. Even in systems where low-severity fires predominate, extreme weather conditions often cause large or intense fires to burn just as intensely in treated areas as untreated ones.[11] Of course, in ecosystems where fire is driven primarily by dry fuels rather than by the continuity or density of fuels, thinning trees or shrubs cannot significantly affect fire behavior at all. Given the cost, the variable effectiveness, and the negative ecological impacts of mechanical thinning, some might question whether it is a useful fire management tool at all.

But mechanical fuel reduction can be extremely useful when applied within very specific parameters. Mechanical treatments used directly around homes and other structures in the wildland-urban interface are often the most convenient, economically efficient, and ecologically benign way to protect wildland-urban interface communities from fire. Clearing vegetation within 65 to 130 feet of structures is one of the most effective ways to prevent these buildings from burning.[12] Another effective method is to make the structures themselves as flame resistant as possible. While this is rarely recognized as a fire management technique, it is virtually the only way to protect structures from being ignited by windborne embers, which frequently cause ignitions miles from the fire itself. Both of these points are important enough to bear repeating: the most effective way to prevent homes in the wildland-urban interface from igniting is by making buildings themselves as fire resistant as possible and by clearing vegetation directly surrounding buildings. With this safety margin in place, the only fire that will ignite a home is the kind of large, intense fire that will not be stopped by small-scale fuel treatments or fire suppression efforts. When landowners fail to remove vegetation around their homes and other buildings, the wildland-urban interface actually feeds fires. As we saw with the Cerro

Figure 9. Firefighters from the Fort Huachuca Fire Station No. 2 light a prescribed fire on the Fort Huachuca Military Reservation in southern Arizona. Photo by Guy McPherson.

Grande Fire, some low-burning fires become large and intense only when they hit the dense fuels of residential areas.

Because of concerns about the effectiveness of fire suppression and mechanical fuel reduction, and the corollary damage these techniques can cause, the use of fire to reduce or change the structure of available fuels is increasingly popular with both land managers and the public (see figures 9–11). Fire use includes both prescribed fires and lightning-ignited fires. Fire historian Stephen Pyne notes, "Earlier, fire was considered controlled only if it burned within the parameter of firelines or fuelbreaks; now fire is considered controlled if it burns within the conditions established by a prescription."[13] Fire is often a practical management tool in

Figure 10.   A Fort Huachuca prescribed fire burns through thick, ungrazed grassland and scattered mesquite shrubs. Photo by Guy McPherson.

rough terrain, in isolated areas, and where other tools are impractical, expensive, or even impossible to use. The use of fire can also be combined with mechanical fuel-reduction techniques to burn slash piles after thinning or to maintain an open, grassy understory that has been produced by previous management efforts.

The use of fire is attractive to many land managers and members of the public in part because fire was historically very common in most North American ecosystems and is therefore viewed as "natural." Fire lacks some of the negative associations of other vegetation-management tools, including the use of herbicides, bulldozers, chainsaws, and livestock. The use of fire, whether prescribed or naturally ignited, can be not only benign but also beneficial from an ecological perspective, especially if its management fits with historical fire regimes. Fire use is often relatively inexpensive compared to fire suppression or mechanical thinning, and

Figure 11.   Grass carries prescribed fire across semidesert grassland at Fort Huachuca, Arizona. Photo by Guy McPherson.

it can treat large areas efficiently. This is especially true when the fire is lightning-ignited but burns within prescribed parameters.

Nonetheless, some opposition to fire use still exists, perhaps largely because fire is seen as such a threat to life and property. In the 1980s, a number of researchers examining public perceptions of fire management found "considerable support" for the use of fire in forest management. Most of those surveyed recognized that fires could have both beneficial and harmful effects, but the support for letting low-intensity fires burn was much lower than for prescribed fire. Those who disagreed with either or both practices cited concerns that fires would get out of control or might damage ecosystems.[14] A similar study found that most of those surveyed supported suppression at all costs when private property or high-value timber was at risk.[15] More recent studies have found continued resistance to the idea of letting fires burn, although members of the public are understandably more likely to support fire use when the associated risks to private property, air quality, forests, and recreation are expected to be low.[16] The public is becoming more aware of the benefits of fire use, but much more education is needed. If we expect members of the public to help build meaningful fire management strategies, we must also help them gain a full and accurate understanding of the benefits and risks of fire.

Despite the inherent risks, manipulating fire frequency and intensity can be effective both in reducing the risk to life and property and in maintaining the biological diversity of native organisms. Fire plays a unique and important role in many ecosystems, one that substitutes like mechanical thinning cannot always fulfill. Countless North American species evolved with fire and are in some sense fire-dependent. For example, the cones of jack pine and lodgepole pine require intense heat to release their seeds.[17] Without recurrent fires, these trees would not propagate or spread. Many other common tree species are intolerant of shade, so they depend on fire to open up the canopy at regular intervals. Low-intensity fires can enrich forest soils by recycling the nutrients in dead fuels and other organic materials on the forest floor. This increases soil pH, stimulates nitrification (the conversion of nitrogen to a form available to plants), and increases the availability of many minerals in the soil.[18] Where fires are frequent, they can thin some tree and shrub species, help maintain a grassy substratum, and generally decrease the incidence of intense crown fires. Few of these positive effects can be replicated by mechanical thinning.

However, the effective and beneficial use of fire as a management tool depends on detailed knowledge of the fire ecology of a particular system, including not only fire frequency but also the patterns of fire intensity and seasonality with which native species evolved. To use fire to reduce the overall fire hazard to populated areas or to maintain biological diversity, managers need a great deal of information about the effects of fire on individual species, populations, and collections of species across large areas. Fire use demands that managers possess superb skills in ecology, watershed function, adaptive long-term management, smoke management, and public relations.

As with any management tool, there are several drawbacks to the use of fire. First, the effects of fire are by no means permanent. Many species, such as New Mexico locust and quaking aspen, resprout vigorously after fire,[19] so postfire decreases in density can be short-lived. A burned site may even be more densely vegetated a few years after a fire if it is dominated by species that resprout or reproduce quickly. The only comprehensive review of literature to date, by Paulo Fernandez and Herminio Botelho, concludes that, because of the rapid accumulation of fuels in forests, prescribed fires are effective in reducing fire hazard for a period of only two to four years.[20] This review indicates that fire hazards can be reduced only by conducting prescribed fires at very frequent intervals. Fernandez and Botelho go on to suggest that prescribed fires are likely to yield effective and reliable results only where extreme weather conditions are unlikely, a rare situation across most of the western United States. Conducting prescribed fires at such short intervals would be inconsistent with historical fire-return intervals in most systems. We would expect such frequent fires to negatively affect soils, aquatic systems, air quality, recreational opportunities, and biological diversity. Where fire is used, very narrow parameters for weather, fuel moisture, and other factors must be met. Several years may pass while fire managers wait for an appropriate time to burn a particular area. As a result, it is often difficult to burn enough area at a sufficient frequency to actually affect fire behavior.

It is important to remember that one fire does not necessarily prevent or mitigate later fires. We often assume that a burned area is unlikely to reignite, but this is not necessarily true. Many fires kill trees but do not burn them completely, with the result that a great deal of dead, dry fuel is left on the landscape. The Great Idaho Fires of 1910 provide a telling example. Trees killed by the 1910 fire also fueled major fires in 1919, 1926, 1929, and 1934, indicating that even the very high-intensity fires of 1910 did not prevent later fires and probably did not reduce their severity.[21]

As we have seen, the effects of contemporary fires are often unlike those of the fires with which species evolved and communities developed. In addition to changes in the mix of species and wildland fuels, habitat fragmentation is a serious issue when fire is used to meet management goals. As noted earlier, former metapopulations of many plant and animal species no longer exist as sources for small, isolated populations. For example, fire-induced floods lead to substantial movement of sediment, and this sediment chokes small streams and extirpates local populations of aquatic organisms.[22] Such events have undoubtedly occurred with some regularity during the last ten thousand years and therefore might be considered natural. Only during the last century has human-caused fragmentation eliminated all potential paths of re-occupation for many such sites. Populations of aquatic organisms that were previously interconnected are now isolated and susceptible to being wiped out by fire-caused sediment movement. As a result, many scientists believe that fires should not be used when populations of rare species are nearby. Unfortunately, the list of threatened and endangered species is constantly growing. Such populations are nearly always in the vicinity of a fire, so most fires are potentially disastrous from this perspective.

The establishment and spread of nonnative species into many fire-prone environments creates even further complexity. Nonnative species, now common throughout the western United States and the world, reduce biological diversity and alter fire regimes in many systems. The combination of long-term fire suppression and introduction of nonnative species has changed species composition and fuels in many, if not most, of the world's ecosystems. As the example of western deserts demonstrated in chapter 2, fires in these altered ecosystems may benefit nonnative species to a greater extent than they benefit native species, thereby threatening rather than enhancing biological diversity. If a nonnative species is prevalent or potentially prevalent on a particular site, land managers should exercise considerable caution when attempting to reintroduce fire.

These problems illustrate the complexity associated with reintroducing fire into ecosystems, even systems in which fire was once a dominant ecological force. These problems also demonstrate the site-specific, local nature of management decisions, and show that fire is neither panacea nor plague. Even in wildlands where fire was once common or where intense crown fires were once normal, fire may now seriously threaten the survival of species that have already been decimated by human activities.

Today, fire management is often treated as if it were equivalent to fuel management, as if we could prevent fires instead of fighting them or living with them. Fire management is strongly influenced by fuel management, but both are also affected by ignition sources, land use, species composition, and weather and climate patterns. Managing fuels is not enough.

Although individual fire managers have no direct control over natural ignition sources and weather, they must consider the relative roles of weather, fuels, and ignition. Ongoing human activities are continuing to fragment ecosystems and introduce nonnative species, which may make it difficult for managers to either reduce the fire hazard or reintroduce fire on a particular site. Indeed, the future consequences of fragmentation and biological invasions will probably limit fire-management options more than any other factor.

Removing dense or laddered fuels may temporarily substitute surface fires of lower intensity for stand-replacing crown fires. Sufficiently aggressive pruning and thinning may substantially reduce overall fire hazard in some ecosystems. While such treatments are expensive and temporary, they may be justified by the significant threat of fire and the associated values at risk. Removing fuels within one hundred or so feet of structures in the wildland-urban interface is particularly effective, as are efforts to make structures more fire-resistant. Ultimately, however, weather and other factors drive fire regimes in many and perhaps most western forests, and the weather increasingly favors fire. Those who see mechanical fuel reduction as a panacea for fire-prone forests should bear in mind the final sentence of Richard Feynman's report on the tragic explosion of the *Challenger* space shuttle: "For a successful technology, reality must take precedence over public relations, for Nature cannot be fooled."[23] The evidence shows that where fire regimes are not primarily driven by fuel density, no amount of political grandstanding will make mechanical thinning prevent fires.

SIX

Policy Solutions

In the policy arena, wildland fire is often depicted as a political problem, a conflict between ideologies or competing land uses. Agency officials accuse environmental activists of blocking what they see as urgently necessary thinning projects. Environmental activists accuse agencies of undermining environmental policies under the guise of fire prevention. Residents of the wildland-urban interface blame either or both, or they point to irresponsible logging and grazing practices. Urban residents wonder why exurban dwellers are still building wood homes in fire-prone forests. Recreationists get angry when fire restrictions interfere with their enjoyment of public lands. Logging companies see fire use as a waste of good lumber.

But from a policy perspective, these conflicts are more or less peripheral. They are built on sound bites, caricatures, and a healthy (if eternal) debate over the proper uses of public lands. They need not preclude the development of a national fire policy, because such a policy cannot and should not resolve these conflicts. We need a policy that will instead establish the overall goals and priorities of fire management for existing uses and, in doing so, provide land-management agencies with a realistic legal and political framework for decision making. At the same time, as members of a democratic society, we are challenged to build policies that make democracy work better and that permit and encourage citizen involvement in decision making

Therefore, we need a policy that will resolve the spatial problem inherent in fire management: the need to simultaneously set national standards and allow for a great deal of variation over time and across different landscapes. It must establish national standards for managing fire, because the management of federal lands affects the whole country; not all policy decisions can be left entirely in the hands of local-level managers. At the same time, we know that fire ecology is so complex and variable that no monolithic national policy could possibly develop a management prescription that works in all ecosystems. The responsibility for deciding when, where, and how to apply the basic tools of fire management must ultimately lie with the local experts: the agency personnel, scientists, residents, and others who know the ecosystems in which they live and work.

The policies of the Healthy Forests Initiative are, paradoxically, both too general and too limited in scope to meet this problem. As we have seen, they underestimate variability in fire regimes by assuming that all fires are controlled primarily by fuel density. Moreover, this initiative fails to establish an overarching framework of priorities and standards that can be applied to all federal lands. Fire ecologists Jerry Franklin and James Agee join many others when they argue that, as a result, the United States today has "no comprehensive policy to deal with fire and fuels and few indications that such a policy is in development."[1] To see why this is more or less true, we must take a brief look at the very recent history of fire policy in the United States.

FIRE POLICY, 1995–2005

The year 2000 was a landmark year for wildland fire. That year, 122,827 fires burned a record 8,422,237 acres across the country.[2] The Cerro Grande Fire burned parts of the town of Cerro Grande, New Mexico, and damaged the Los Alamos National Laboratory. That fire spread across some 47,650 acres and destroyed 235 structures.[3] These numbers, in combination with the usual nightly news videos of advancing flames and charred forests, gave the public a clear impression of fires burning out of control. In the wake of the Cerro Grande fire in particular, there was widespread criticism of federal agencies' fire-management practices. All in all, the 2000 fire season made policy makers want to do—or at least to be seen as doing—something about fire.

Interestingly enough, something already had been done. Following an earlier news-making fire season, in 1994, President Bill Clinton had directed Secretary of Agriculture Dan Glickman and Secretary of Interior Bruce Babbitt to form a wildland fire working group.[4] They sought, ultimately, to create a national policy for coping with wildland fire. The 1994 season was not particularly extreme in terms of the number of fires or acres burned, but it was unusually deadly. Thirty-five wildland firefighters were killed, more than in any other season since the catastrophic Great Idaho Fires of 1910.[5] The need to ensure firefighter safety became a significant theme in the group's work.

The resulting document, released in 1995, was the Federal Wildland Fire Management Policy and Program (henceforth, the 1995 Federal Fire Policy). It aimed to provide clear, cohesive direction in fire management to the relevant federal agencies by establishing guiding principles for all fire-management agencies. These guiding principles emphasized the primacy of firefighter safety, the role of fire as a necessary and fundamental ecological process, and the need to make fire protection economically viable. The working group emphasized fire preparedness (including the reduction of heavy fuel loads in some areas) over fire suppression.[6]

In their efforts to build a usable national fire policy, the 1995 working group moved beyond establishing guiding principles, although this alone was a major step forward. They also outlined a plan for implementation. They directed land-management agencies (including the USDA Forest Service, National Park Service, and Bureau of Land Management) to develop a fire management plan for every burnable area of federal land. These plans, often referred to as FMPs, were to outline a strategy for the use, prevention, and suppression of fire throughout the area being managed. The fire management plans were to "be consistent with firefighter and public safety, values to be protected, and public health issues" and "address all potential wildland fire occurrences and include the full range of fire management actions."[7]

Agencies were somewhat slow to follow these recommendations, in part because funding and oversight were minimal. Fire management plans are subject to review under the National Environmental Policy Act, so agencies must also take the time to examine potential environmental effects of their plans and present them to the public for comment. Many areas lacked fire management plans even by 2001, when the working group reconvened for a review and update of the policy.[8] Today, however, federal agencies increasingly see fire management planning as a major part of their mandate, and many forests and parks are enthusiastically designing fire management plans according to the Federal Fire Policy's recommendations. By 2006, nearly all fire management plans had been completed.[9]

The 1995 Federal Fire Policy was impressive in many respects. It addressed the full range of fire management strategies, and it recognized fire's vital role in wildland ecosystems. Perhaps most important, it tackled the spatial problem inherent in fire management: the need to simultaneously set national standards and allow for a great deal of variation over time and across different landscapes. The guiding principles neatly avoided ideological debates by focusing on matters of broad consensus: the need to prioritize public and firefighter safety, the beneficial role of fire in ecosystems, and the importance of using all available tools to manage fire effectively. This approach is crucial to the policy's practical usefulness. Again, the appropriate goal of a national fire policy is to create priorities, structures, and institutions for managing fire. It should not aim to resolve the problem of competing interests and uses. In this sense, the Federal Fire Policy was highly successful.

We cannot say that fire management plans have proven to be immune to ideology, but their format and content make a great deal of sense. They incorporate the policy's guidelines and apply them to local conditions. Based on existing land uses, landscapes, and fire regimes, fire management plans map and define fire management units. Agencies prescribe different kinds of fire management for different units: a unit in the wildland-urban interface can be treated differently than a wilderness area; an area of mixed-conifer forest might have a management plan different from that of a nearby ponderosa pine stand. Every step of the plan, from the definition of management units to the specific strategies for suppression, thinning, or fire use, is subject to public review and comment, and agencies are generally encouraged to make the planning process as collaborative as possible.[10]

Nonetheless, few people not directly involved in forest planning have ever heard of the 1995 Federal Fire Policy. Its implementation had barely even begun when the 2000 fire season sparked a new round of fervent policy making, and the policies that followed almost completely overshadowed it. First, the 1995 working group reassembled to produce a "review and update" document that was released in 2001, and that officially replaced the 1995 policy (henceforth, the 2001 Federal Fire Policy). Although the group changed the guiding principles very little, they found that implementation was inconsistent and incomplete, and that the failure of federal agencies to fully implement the 1995 policy had seriously undermined its effectiveness.[11]

As the 2000 season wore on, President Clinton looked again to Babbitt and Glickman, requesting that they prepare an analysis of the nation's overall wildland fire situation.[12] This report formed the basis of the National Fire Plan, which we mentioned briefly in chapter 3. The National Fire Plan represented yet another attempt to clarify priorities and set guidelines for fire management, but its primary result was the allocation of funding for various fire-management activities, especially for fire suppression. Annual budget appropriations for fire management to the Forest Service and the Department of Interior were increased from $1.6 billion in fiscal year 2000 (the year before the National Fire Plan went into effect) to $2.2 billion in 2001 and to $3.2 billion in 2005. Bizarrely, and in complete contradiction to the 1995 and 2001 Federal Fire Policies, fully half of that budget was devoted to fire-suppression activities.[13]

The policy situation became even more complex in 2001, when a new plan was developed to implement the ideas of the National Fire Plan. This document was generally referred to as the "10-Year Comprehensive Strategy." It placed heavy emphasis on fire suppression and prevention and contradicted already existing policies. Shortly afterward, the various policies of the Healthy Forests Initiative were added to the list. This hodgepodge of confusing and often contradictory policies helps explain why, in spite of the evidence-backed 1995 and 2001 Federal Fire Policies, we do not see a cohesive national fire policy today. The Federal Fire Policies, which showed a great deal of promise but which were not swiftly implemented, were effectively drowned in the new initiatives.

Today, the result must be seen as an overt policy failure. More than a decade after the 1995 Federal Fire Policy was first released, the federal government is increasing the amount of money it spends on fire management nearly every year, without seeming to get any closer to its stated goals. Fire management plans, the most promising element of recent policies, are nearly completed, but the Government Accountability Office warns that updating or modifying these plans may prove difficult.[14] In any case, the federal fire policy was produced not by the legislative branch but by the executive branch of a past administration, making its mandate tenuous at best. After more than ten years of policy change, a practical issue of social and scientific management has been turned into a political minefield.

When policy fails, citizens are the ones who suffer. The original problems, the ones the policy was supposed to address, stand uncorrected, while the unintended consequences multiply and spread. In the case of the failed fire policies discussed here, our democratic systems for governance have themselves been victims.

We have described a series of social and political problems caused by failed fire policies. These include massive disruptions to the normal budget process, an increasing concentration of decision-making power in land-management agencies, the creation of perverse subsidies for development in the wildland-urban interface, and a concomitant undermining of processes for citizen participation in public-lands planning.

The previous chapter showed how fire management can be improved by focusing management efforts on the maintenance and restoration of biological diversity on public lands and explicitly placing a higher priority on protecting human life than on protecting property from fire. We showed that these changes can be accomplished through the careful, informed, and above all discriminate application of different management techniques, including fire suppression, mechanical thinning, and fire use. We suggested using mechanical thinning to remove small-diameter fuels primarily within 130 feet of structures and, in some cases, in forests where fire occurrence and behavior are strongly influenced by fuel density and structure. We also suggested that, to restore socially and ecologically acceptable fire regimes in some systems, public land managers must increase their use of (and tolerance of) intentionally and naturally ignited fire. When used with attention to historical regimes, fire is certainly among the most valuable tools available to managers. Finally, we argued that fire-suppression efforts are best limited to certain kinds of fires in the wildland-urban interface, and that some very intense fires simply cannot be suppressed successfully. We do not support, and do not believe the public should allow, the threat to firefighters' lives and the waste of public funds associated with fighting unmanageable fires for political reasons. While some would find limiting suppression efforts counterintuitive or dangerous, we argue that such actions will not increase the fire risk to human life and property *if* this change in suppression policy is made in concert with the other recommended changes in thinning and fire use practices.

We designed these guidelines to mitigate the negative ecological impacts of fire management. They also represent a good first step toward mitigating the social damage caused by past and present fire-management policies and practices. Managing public lands to protect biological diversity, rather than to protect the private real estate on their borders, will reduce the massive public subsidies that currently promote development in the wildland-urban interface. The annual apportionment that emergency fire suppression takes from management agencies' budgets will be reduced if unnecessary suppression efforts are halted or minimized. Where management focuses more on protecting local systems than on applying a broad national policy of fire suppression or forest thinning, we may also expect to see more opportunities for citizen participation in managing public lands. However, addressing long-standing problems with fire management will also require broader and more fundamental policy changes.

Policy analysis has traditionally focused on efficiency; like the architects and proponents of the Healthy Forests policies, experts have generally evaluated policies on their efficiency in reaching their goals. More recently, a few policy analysts (most notably

Helen Ingram, Steven Rathgeb Smith, and Marc Landy) have suggested that we judge policies by a more meaningful standard. They argue that we should pay more attention to the overall effect policies have on citizens and especially on citizenship and democracy. In doing so, these analysts have begun to pay long-overdue attention to the effects that different policies have on the democratic processes of deliberation and self-governance.[15]

These researchers imagine an optimistic future for policy. They describe the potential for policy to "empower, enlighten, and engage citizens in the process of self-government."[16] What would such policies look like? Among other traits, they would certainly have clear goals and transparent operations. In doing so, they would limit the exercise of bureaucratic discretion and make it easier for both citizens and their elected representatives to oversee the policy process.[17] Where policy goals and implementation are obscured by unnecessary complexity, citizens are unmotivated or even unable to participate. It matters little to the public whether this obfuscation is intentional or unintentional.

Smith and Ingram argue that "an essential aspect of a well-functioning democracy is the capacity for deliberation and discussion."[18] It is clear that this is what the founding fathers had in mind, and it fits very well with our ideal picture of democratic processes. It follows, then, that democratic policies should support and encourage this capacity. Today, the policies that we point to as accommodating citizen participation, such as the Administrative Procedures Act, the Appeals Reform Act, and the National Environmental Policy Act, do not really encourage democratic participation or deliberation. At best, they merely permit such expressions of citizenship; at worst, they make the participation process so convoluted, so burdensome, or so constricted that substantive citizen involvement is virtually impossible. Even though federal agencies often go beyond the minimum requirements to promote citizen participation in decision making, the formal hearing, appeal, and litigation processes are unfriendly and difficult for the average citizen to navigate. With the exception of mass-mailing campaigns organized by citizen groups, participation in these processes is generally low; participation also tends to focus on ideological conflicts and to be more or less limited to middle- and upper-class citizens.[19]

When we call for policies that encourage democratic deliberation and public participation, we are not talking only about the most basic rights of access to information and protection in the courts. These rights must, of course, be protected, and they must be restored where they have already been eroded by legislative, judicial, and administrative changes. Policies like the National Environmental Policy Act are necessary to democratic policy and management, but they are not sufficient. In moving beyond policy efficiency and creating policies that support democratic processes, transparency and legal access will not be sufficient. The question of how to craft such policies is both deep and broad, and it may require substantial changes to the ways we think about and practice governance. In order to approach this question in this limited forum, we return once again to the specific realm of fire management on public lands. How can we conceive of a democratically and ecologically sound fire policy?

NEW DIRECTIONS

As we build fire policy for our public lands, we face both a need for clear management priorities at the national level and a need for flexibility and ecosystem-specific decision-making at the local level. The best policy tool for living effectively and safely with wildland fire is a familiar one. We suggest that a devolution, or decentralization, of fire management is necessary for resolving the inherent spatial conundrum we describe here. This idea is certainly not new; many policy thinkers over many years have argued that decentralization is the key to encouraging active citizenship.[20]

Landy, one of the key researchers examining the effects of policies on democratic processes, is also a major proponent of political decentralization. He argues, "The centralization of policy direction at the national level is inimical to citizenship because it deprives local civic forums of weighty matters to deliberate about. Citizens are unlikely to expend the painstaking effort civic participation requires if they have the sense that all the really important decisions are being made in Washington. Centralization also encourages citizens to be irresponsible. If the funds to handle a particular problem are attainable from Washington, then there is no reason for the local citizenry to fix it themselves."[21] In the context of fire management, the legitimacy of this argument is particularly clear. When federal policies specify fire suppression or prevention as a national norm, the public has no reason to invest time and effort in deliberating over the details of on-the-ground management. Similarly, when national policies support one group of land- or resource-users over others, the public is distanced from the very important question of balancing competing uses. When the federal budget allows for unlimited spending on fire suppression, there is no reason for those living in the wildland-urban interface to manage fuels on their property or to help make decisions about how best to manage wildland fires.

It is extremely important that any attempt at political decentralization, unlike many past efforts, not weaken existing regulatory policies. At the national level, regulatory environmental policies such as the Endangered Species Act and the National Environmental Policy Act have been major targets of recent efforts at deregulation. As we saw in chapter 5, arguments for the Healthy Forests policies return frequently to the idea that these policies inhibit appropriate fire-management practices. The need for flexibility in fire management at the local level should not be used as an excuse to undermine regulatory, environmental, and other federal policies that protect both citizens and ecosystems. Instead, strong federal standards and guidelines are needed to provide a framework for fire management.

The development of a streamlined and logically consistent national-level fire policy must be the first step in reworking fire-management practices. This policy should explicitly replace all previously existing policies and should emphasize the priorities we have already outlined, those of protecting firefighter and public safety, of maintaining and restoring biological diversity and the processes with which species evolved, and of protecting property from fire, when possible, through the best available method: managing fuels directly adjacent to flammable structures. Because most of these priorities are addressed in the Federal Fire Policy, we suggest building on the 2001 plan. A new revision should be centered on a realistic view of fire management and must not fall back on antiquated and dysfunctional ideas of preventing or controlling large fires. More important, however, this policy must be recognized as *the* Federal Fire Policy. Ideally, it would be a legislative effort; at the very least, it should include a mechanism for oversight and a clear and specific request for funding.

The second step in revamping fire policy must be to bring budget allocation and staffing in line with stated policy priorities. It is important to acknowledge on paper the harmful effects of fire suppression and to give priority to other, more useful management tools. However, such rhetoric is practically useless when contradicted by agency budgets. A funding process that allows unlimited emergency allocations to firefighting efforts provides a strong incentive for fire suppression, regardless of statements about other policy priorities. As politically risky as such a decision will be, policy makers must find a more reasonable way to allocate fire-management funds. As long as budgets for homeowner education, fuel treatment, and especially the reintroduction of fire are regularly reappropriated to allow unlimited funds for suppression, there can be no hope for ecologically based management of fire on our public lands. Given the current state of the federal budget, we assume that funds for these important activities will have to come from a redistribution of agency funds rather than from a large allocation of new funds for proactive fire management. Eliminating the National Fire Plan allocations, with their clear emphasis on suppression activities, will be a good start.

Third and finally, national fire policy must have a clear and effective mechanism for placing local fire-management decisions in the hands of local managers and citizens. In this effort, again, the Federal Fire Policy is more than halfway there. Now that most federal lands have fire management plans, they need the money and support to implement them. This limited form of delegation should not be confused with a total allocation of decision-making power to local officials. In a federalist system, natural resource policy developed at local levels tends to emphasize economic rather than environmental concerns.[22] This is both understandable and, in many cases, appropriate, but local economic interests should not drive the management of our public-owned lands. Because of the interests all citizens have in these lands, it would be inappropriate to allow a complete localization of decision making without maintaining enforceable national standards for the care and preservation of those resources.

Many researchers and theorists have observed an ongoing concentration of power in bureaucratic agencies and have argued for a return to governance at local scales, which are assumed to be more accessible to citizens. While democratic governance is an important priority, we must not assume that a move to local-level management will automatically renew democratic processes and create more egalitarian fire management. Fire management plans are currently subject to review under the National Environmental Policy Act, which provides for a minimal level of citizen participation. As changes to regulations undermine that policy's application to fire management decisions, however, fewer decisions will be open to even the very narrow democratic processes it protects. The first step in protecting democratic participation is to restore and strengthen the laws that allow for it.

But further changes must follow, and it is difficult to conceptualize how these changes could be incorporated into formal policy. We must find a way to create opportunities for real involvement, in which the public is not merely allowed to comment on management decisions but is actively encouraged to deliberate on and create meaningful change. Federal agencies must take an active role in this process; senior officials in particular should learn to encourage substantive public debate. To do so, they will have to resist the temptation to fuel ideological debates and land-use conflicts. Instead, they should seize the opportunity to present management problems to the public in a way that encourages thoughtful deliberation. Members of the general public, in turn, will have to find a way to move beyond irresolvable ideological disagreements and to enter into meaningful political dialog with their neighbors.

In the first chapter of this book, we described how the idea of a modern wildland-fire crisis has been built on misinformation and miscommunication. We have since explored how fire regimes vary across time and space and examined the causes of their variability. We looked at a few failed attempts to control wildland fire, and we proposed our own solutions for learning to live with it and use it as a tool. Our recommendations are fairly straightforward: fire management should prioritize human safety, biological diversity, and, to a lesser extent, private property. In the long term, the first two are best protected by the careful reintroduction of fire as a normal part of wildland ecosystems; we should protect the latter through relatively limited management efforts directly in the wildland-urban interface. Fire suppression must be recognized as an inherently harmful effort and can no longer be seen as a default function of federal agencies. Where agencies choose to fight fire, they should have good reason to do so and should explicitly recognize the potential negative effects. Finally, a cohesive federal fire policy is needed to guide, fund, and oversee these efforts.

We see a generally positive trend in fire management today. Fire is now viewed as an inevitable event in most North American wildlands. Managers, progressive policy-makers, and the informed public often see it as a positive, even necessary process. The role of fire in maintaining biodiversity is likewise well recognized. Fire is widely perceived to be an appropriate tool for reducing fire hazard. Although this idea is often applied too broadly by policy makers, managers, and the public, it nonetheless represents a marked improvement over the earlier assumption that all fires should be suppressed. Today there is some danger that the dangerous and harmful fire-suppression policies will be replaced by an equally flawed single-minded focus on fire prevention through fuel reduction. In our interactions with fire managers and the public, however, we have found their perceptions of fire ecology to be increasingly nuanced and increasingly accurate.

NOTES

INTRODUCTION

1. Alvin Toffler, *Powershift: Knowledge, Wealth, and Violence at the Edge of the Twenty-first Century* (New York: Bantam Books, 1990), xvii.

2. Wildland fire issues are of course not limited to the western United States, but this book focuses almost exclusively on that region. The West is generally seen as the center of the national fire crisis, at least in part because of its juxtaposition of large areas of fire-prone public land with rapidly expanding private development projects.

3. James Brian McPherson, *Journalism at the End of the American Century, 1965–Present* (Westport, CT: Praeger, 2006).

1. WILDLAND FIRE IN THE WEST

1. Technically, of course, fires lit by arsonists can be as useful as lightning fires, but land management agencies are understandably loath to let such fires burn.

2. A. S. Leopold, S. A. Cain, D. M. Cottam, I. N. Gabrielson, and T. L. Kimball, "Wildlife Management in the National Parks," Transactions

of the North American Wildlife and Natural Resources Conference 28 (1963): 28–45.

3. A number of names have been applied to this technique, including "let-burn" and, more recently, "wildland fire use."

4. T. Schoennagel, M. G. Turner, and W. H. Romme, "The influence of Fire Interval and Serotiny on Postfire Lodgepole Pine Density in Yellowstone National Park," *Ecology* 84, no. 11 (2003): 2967–2978.

5. National Park Service, U.S. Department of the Interior, "Yellowstone National Park: Wildland Fire in Yellowstone," n.d., pp. 1–2, available from National Park Service, www.nps.gov/yell/naturescience/wildlandfire.htm.

6. David Carle, *Burning Questions: America's Fight with Nature's Fire* (Westport, CT: Praeger, 2002).

7. N. L. Christensen, J. K. Agee, P. F. Brussard, J. Hughes, D. H. Knight, G. W. Minshall, J. M. Peek, S. J. Pyne, F. J., Swanson, J. W. Thomas, S. Wells, S. E. Williams, and H. A. Wright, "Interpreting the Yellowstone Fires of 1988: Ecosystem Responses and Management Implications," *BioScience* 39, no. 10 (1989): 678–685.

8. Carle, *Burning Questions.*

9. National Park Service, "Yellowstone National Park," p. 2.

10. Stephen J. Pyne, *World Fire* (New York: Holt, 1995), p. 256.

11. Yvonne Baskin, "Yellowstone Fires: A Decade Later," *BioScience* 49, no. 2 (1999): 93–97.

12. Carle, *Burning Questions.*

13. Baskin, "Yellowstone Fires."

14. Monica G. Turner, William H. Romme, and Donald B. Tinker, "Surprises and Lessons from the 1988 Yellowstone Fires," *Frontiers in Ecology and the Environment* 1, no. 7 (2003): 351–358.

15. Carle, *Burning Questions.*

16. Stephen J. Pyne, Patricia L. Andrews, and Richard D. Laven, *Introduction to Wildland Fire*, 2nd ed. (New York: John Wiley and Sons, 1996).

17. Orville Daniels, "A Forest Supervisor's Perspective on the Prescribed Natural Fire Program," in *Proceedings of the 17th Tall Timbers Fire*

Ecology Conference: High Intensity Fire in Wildlands, May 18–21, 1989 (Tallahassee, FL: Tall Timbers Research Station, 1989), p. 365

18. William L. Baker, and Donna Ehle, "Uncertainty in Surface-Fire History: The Case of Ponderosa Pine Forests in the Western United States," *Canadian Journal of Forest Research* 31 (2001): 1205–1226.

19. Guy R. McPherson and Jake F. Weltzin, *Disturbance and Climate Change in United States/Mexico Borderland Plant Communities: A State-of-the-Knowledge Review*, General Technical Report RMRS-GTR-50 (Fort Collins, CO: U.S. Department of Agriculture, Forest Service, Rocky Mountain Research Station, 2000); Guy R. McPherson, "The Role of Fire in Desert Grasslands," in *The Desert Grassland*, ed. Mitchel P. McClaran and Thomas R. Van Devender, pp. 130–151 (Tucson: University of Arizona Press, 1995).

20. Pyne et al., *Introduction to Wildland Fire*.

21. Stephen J. Pyne, *Fire in America: A Cultural History of Wildland and Rural Fire* (Seattle: University of Washington Press, 1997).

22. Henry A. Wright and Arthur W. Bailey, *Fire Ecology: United States and Southern Canada* (New York: John Wiley, 1982).

23. Paulo M. Fernandes and Herminio S. Botelho, "A Review of Prescribed Burning Effectiveness in Fire Hazard Reduction," *International Journal of Wildland Fire* 12 (2003): 117–128.

24. Barbara Goodrich Phillips and Debra Crisp, "Dalmatian Toadflax, an Invasive Exotic Noxious Weed, Threatens Flagstaff Pennyroyal Community Following Prescribed Fire," in *Southwestern Rare and Endangered Plants: Proceedings of the 3rd Conference; 2000 September 25–28*, Joyce Maschinski and Louella Holter, technical coordinators, Proceedings RMRS-P-23 (Fort Collins, CO: U.S. Department of Agriculture, Forest Service, Rocky Mountain Research Station, 2001), pp. 200–205.

25. William J. Bond and Jon E. Keeley, "Fire as a Global 'Herbivore': The Ecology and Evolution of Flammable Ecosystems," *Trends in Ecology and Evolution* 20, no. 7 (2005): 387–394; A. M. Gill, R. A. Bradstock, and J. E. Williams, "Fire Regimes and Biodiversity: Legacy and Vision," in *Flammable Australia: The Fire Regimes and Biodiversity of a Continent*, ed. R. A. Bradstock, J. E. Williams, and M. A. Gill, pp. 449–46 (Cambridge,

U.K.: Cambridge University Press, 2002); Allen A. Steuter and Guy R. McPherson, "Fire as a Physical Stress," in *Wildland Plants: Physiological Ecology and Developmental Morphology,* ed. Donald J. Bedunah and Ronald E. Sosebee, pp. 550–579 (Denver, CO: Society for Range Management, 1995).

26. General Accounting Office, *Wildland Fire Management: Additional Actions Required to Better Identify and Prioritize Lands Needing Fuels Reduction,* GAO-03–805 (Washington, DC, 2003), p. 4.

27. Jolie Pollet and Philip N. Omi, "Effect of Thinning and Prescribed Burning on Crown Fire Severity in Ponderosa Pine Forests," *International Journal of Wildland Fire* 11 (2002): 1–10.

28. Ertugrul Bilgili, "Stand Development and Fire Behavior," *Forest Ecology and Management* 179 (2003): 333–339.

29. As many others have noted, the concept of fire severity is, in and of itself, problematic. Dr. Jon Keeley of the U.S. Geological Survey describes a situation in which the terms *fire intensity* and *fire severity* are incorrectly used to describe the physical and ecological effects of fires. *Fire intensity* is appropriately used to describe the actual energy output of the fire, while *fire severity* refers specifically to "the loss of or change in organic matter above ground and below ground." We find ourselves guilty of using an inappropriately broad interpretation of *fire severity* here, but as Keeley also notes, "ecosystem responses are ultimately what are of most interest to resource managers" (Jon E. Keeley, "Fire Intensity, Fire Severity, and Related Terminology: A Brief Review and Suggested Usage," unpublished article in review for *International Journal of Wildland Fire*, 1). For want of a more suitable term, we follow Schoennagel and others (see n. 30) in using *fire severity* as a shorthand reference to a much broader and more complex set of variables.

30. Tania Schoennagel, Thomas T. Veblen, and William H. Romme, "The Interaction of Fire, Fuels, and Climate across Rocky Mountain Forests," *BioScience* 34, no. 7 (2004): 661–676.

31. T. W. Swetnam and J. L. Betancourt, "Mesoscale Disturbance and Ecological Response to Decadal Climatic Variability in the American Southwest," *Journal of Climate* 11 (1998): 3128–3147; T. Kitzberger,

T. W. Swetnam, and T. T. Veblen, "Inter-Hemispheric Synchrony of Forest Fires and the El Niño–Southern Oscillation," *Global Ecology and Biogeography* 10, no. 3 (2001): 315–326.

32. T. W. Swetnam, C. H. Baisan, and J. M. Kaib, "Forest Fire Histories in the Sky Islands of La Frontera," in *Changing Plant Life of La Frontera: Observations on Vegetation in the United States/Mexico Borderlands*, ed. G. L. Webster and C. J. Bahre, pp. 95–119 (Albuquerque: University of Mexico Press, 2001); T. W. Swetnam, "Fire Histories from Pine-Dominant Forests," in *Proceedings of the Conference: Biodiversity and Management of the Madrean Archipelago II: Connecting Mountain Islands and Desert Seas, May 11–15, 2004, Tucson, AZ*, General Technical Report RMRS-P-36:35–43 (Fort Collins, CO: U.S. Department of Agriculture, Forest Service, Rocky Mountain Research Station, 2005).

33. James E. Lotan, James K. Brown, and Leon F. Neuenschwander, "Role of Fire in Lodgepole Pine Forests," in *Lodgepole Pine: The Species and Its Management: Symposium Proceedings: 1984 May 8–10, Spokane, WA; 1984 May 14–16, Vancouver, B.C.*, ed. and comp. David M. Baumgartner, Richard G. Krebill, James T. Arnott, and Gordon F. Weetman, pp. 133–152 (Pullman, WA: Washington State University, Cooperative Extension, 1985).

34. Schoennagel et al., "Interaction of Fuels, Fire, and Climate"; E. A. Johnson, *Fire and Vegetation Dynamics, Studies from the North American Boreal Forest* (Cambridge: Cambridge University Press, 1992); W. C. Bessie and E. A. Johnson, "The Relative Importance of Fuels and Weather on Fire Behavior in Subalpine Forests," *Ecology* 76, no. 3 (1995): 747–762; J. E. Keeley, C. J. Fotheringham, and M. Morias, "Reexamining Fire Suppression Impacts on Brushland Fire Regimes," *Science* 284 (1999): 1829–1832.

35. Schoennagel et al., "Interaction of Fuels, Fire, and Climate."

36. Wright and Bailey, *Fire Ecology*.

37. T. T. Kozlowski and C. E. Ahlgren, eds., *Fire and Ecosystems* (New York: Academic Press, 1974).

38. James Agee, *Fire Ecology of Pacific Northwest Forests* (Washington, DC: Island Press, 1996).

39. Pyne et al., *Introduction to Wildland Fire*.

40. David E. Brown, ed., "Biotic Communities of the American Southwest—United States and Mexico," *Desert Plants* 4, no. 1–4 (1982): 1–342.

41. Peter Z. Fulé, Charles McHugh, Tomas A. Heinlein, and W. Wallace Covington, "Potential Fire Behavior Is Reduced Following Forest Restoration Treatments," in *Ponderosa Pine Ecosystems Restoration and Conservation: Steps toward Stewardship*, comp. Regina K. Vance, C. B. Edminster, W. W. Covington, and J. A. Blake, pp. 28–35, Proceedings RMRS-P-22 (Fort Collins, CO: U.S. Department of Agriculture, Forest Service, Rocky Mountain Research Station, 2001), p. 28.

42. See, for example, Brown, "Biotic Communities of the American Southwest"; Wright and Bailey, *Fire Ecology;* Pollet and Omi, "Effect of Thinning and Prescribed Burning," 1–10.

43. Guy R. McPherson, *Ecology and Management of North American Savannas* (Tucson: University of Arizona Press, 1997).

44. Evolutionary adaptations and fire ecology of ponderosa pine are described in the following sources: Harold Weaver, "Fire and Its Relationship to Ponderosa Pine," *Proceedings of the Tall Timbers Fire Ecology Conference* 7 (1967): 127–149; Charles E. Boldt and James L. Van Deusen, *Silviculture of Ponderosa Pine in the Black Hills: The Status of Our Knowledge*, Research Paper RM-124 (Fort Collins, CO: U.S. Department of Agriculture, Forest Service, Rocky Mountain Forest and Range Experiment Station, 1974); Gilbert H. Schubert, *Silviculture of Southwestern Ponderosa Pine: The Status of Our Knowledge*, Research Paper RM-123 (Fort Collins, CO: U.S. Department of Agriculture, Forest Service, Rocky Mountain Forest and Range Experiment Station, 1974); John H. Dieterich, *Recovery Potential of Fire-Damaged Southwestern Ponderosa Pine*, Research Note RM-379 (Fort Collins, CO: U.S. Department of Agriculture, Forest Service, Rocky Mountain Forest and Range Experiment Station, 1979); Wright and Bailey, *Fire Ecology;* Stephen S. Sackett, *Observations on Natural Regeneration in Ponderosa Pine Following a Prescribed Fire in Arizona*, Research Note RM-435 (Fort Collins, CO: U.S. Department of Agriculture, Forest Service, Rocky Mountain Forest and

Range Experiment Station, 1984); William W. Oliver and Russel A. Ryker, "*Pinus ponderosa* Dougl. Ex Law. Ponderosa Pine," in *Silvics of North America*, vol. 1, *Conifers*, Russell M. Burns and Barbara H. Honkala, technical coordinators, pp. 413–424, Agricultural Handbook 654 (Washington, DC: U.S. Department of Agriculture, Forest Service, 1990).

45. Charles F. Cooper, "Changes in Vegetation, Structure, and Growth of Southwestern Pine Forests since White Settlement," *Ecological Monographs* 30 (1960): 129–164.

46. Wright and Bailey, *Fire Ecology*.

47. A. W. Lindenmuth Jr., *A Survey of Effects of Intentional Burning on Fuels and Timber Stands of Ponderosa Pine in Arizona*, Research Paper RM-54 (Fort Collins, CO: U.S. Department of Agriculture, Forest Service, Rocky Mountain Forest and Range Experiment Station, 1960).

48. Harry E. Brown, "Gambel Oak in West-Central Colorado," *Ecology* 39 (1958): 317–327; Art R. Tiedemann, Warren P. Clary, and Richard J. Barbour, "Underground Systems of Gambel Oak *(Quercus gambelii)* in Central Utah," *American Journal of Botany* 74 (1987): 1065–1071; Peter Z. Fulé, W. Wallace Covington, and Margaret M. Moore, "Determining Reference Conditions for Ecosystem Management of Southwestern Ponderosa Pine Forests," *Ecological Applications* 7 (1997): 895–908.

49. Peyton W. Owston and William I. Stein, *Pseudotsuga* Carr. Douglas-Fir," in *Seeds of Woody Plants in the United States*, ed. C. S. Schopmeyer, pp. 674–683, Agricultural Handbook 450 (Washington, DC: U.S. Department of Agriculture, Forest Service, 1974); Richard K. Hermann and Denis P. Lavender, "*Psedutsuga menziesii* (Mirb.) Franco Douglas-Fir," in *Silvics of North America*, vol. 1, *Conifers*, Russell M. Burns and Barbara H. Honkala, technical coordinators, pp. 527–540, Agricultural Handbook 654 (Washington, DC: U.S. Department of Agriculture, Forest Service, 1990).

50. Gary M. Ahlstrand, "Fire History of a Mixed Conifer Forest in Guadalupe Mountains National Park," in *Proceedings of the Fire History Workshop, October 20–24, 1980, Tucson, AZ*, Marin A. Stokes and John H.

Dieterich, technical coordinators, pp. 4–7, General Technical Report RM-81 (Fort Collins, CO: U.S. Department of Agriculture, Forest Service, Rocky Mountain Research Station, 1980); John Dieterich, "Fire History of Southwestern Mixed Conifer: A Case Study," *Forest Ecology and Management* 6 (1983): 13–31.

51. Wright and Bailey, *Fire Ecology.*

52. Norbert V. De Byle, "The Role of Fire in Aspen Ecology," in *Proceedings: Symposium and Workshop on Wilderness Fire*, James E. Lotan, Bruce M. Kilgore, William C. Fisher, and Robert W. Mutch, technical coordinators, p. 326, General Technical Report INT-182 (Ogden, UT: U.S. Department of Agriculture, Forest Service, Intermountain Forest and Range Experiment Station, 1985); Norbert V. DeByle, Collin D. Bevins, and William C. Fischer, "Wildfire Occurrence in Aspen in the Interior Western United States," *Western Journal of Applied Forestry* 2, no. 3 (1987): 73–76.

53. George A. Schier, John R. Jones, and Robert P. Winokur, "Vegetative Regeneration," in *Aspen: Ecology and Management in the Western United States*, ed. Norbert V. DeByle and Robert P. Winokur, pp. 29–33, General Technical Report RM-119 (Fort Collins, CO: U.S. Department of Agriculture, Forest Service, Rocky Mountain Forest and Range Experiment Station, 1985); D. A. Perala, *Populus tremuloides* Michx. Quaking Aspen," in *Silvics of North America*, vol. 2, *Hardwoods*, Russell M. Burns and Barbara H. Honkala, technical coordinators, pp. 555–569, Agriculture Handbook 654 (Washington, DC: U.S. Department of Agriculture, Forest Service, 1990).

54. Mariette T. Seklecki, Henry D. Grissino-Mayer, and Thomas W. Swetnam, "Fire History and the Possible Role of Apache-Set Fires in the Chiricahua Mountains of Southeastern Arizona," in *Effects of Fire on Madrean Province Ecosystems: A Symposium Proceedings*, Peter F. Ffolliott, Leonard F. DeBano, Malchus B. Baker Jr., Gerald J. Gottfried, Gilberto Solis-Garza, Carleton B. Edminster, Daniel B. Neary, Larry S. Allen, and Robert H. Hamre, technical coordinators, pp. 238–246, General Technical Report RM-GTR-289 (Fort Collins, CO: U.S. Department of Agriculture, Forest Service, Rocky Mountain Forest and Range Experiment Station, 1996).

55. T. W. Swetnam and C. H. Baisan, "Fire Histories of Montane Forests in the Madrean Borderlands," in *Effects of Fire on Madrean Province Ecosystems, a Symposium Proceedings*, Peter F. Ffolliott, Leonard F. DeBano, Malchus B. Baker Jr., Gerald J. Gottfried, Gilberto Solis-Garza, Carleton B. Edminster, Daniel B. Neary, Larry S. Allen, and Robert H. Hamre, technical coordinators, pp. 15–36, General Technical Report RM-GTR-289 (Fort Collins, CO: U.S. Department of Agriculture, Forest Service, Rocky Mountain Forest and Range Experiment Station, 1996).

56. D.J. Shinneman and William L. Baker, "Nonequilibrium Dynamics between Catastrophic Disturbances and Old-Growth Forests in Ponderosa Pine Landscapes of the Black Hills," *Conservation Biology* 11 (1997): 1276–1288; Peter Brown, Merrill R. Kaufmann, and Wayne D. Shepperd, "Long-Term, Landscape Patterns of Past Fire Events in a Montane Ponderosa Pine Forest of Central Colorado," *Landscape Ecology* 14 (1999): 513–532; E. A. Johnson, K. Miyanishi, and S. R.J. Bridge, "Wildfire Regime in the Boreal Forest and the Idea of Suppression and Fuel Buildup," *Conservation Biology* 15 (2001): 1554–1557; Donna Ehle and William L. Baker, "Disturbance and Stand Dynamics in Ponderosa Pine Forests in Rocky Mountain National Park, USA," *Ecological Monographs* 73 (2003): 543–566.

57. See, for example, H. C. Humphries and P. S. Bourgeron, "Environmental Responses of *Pinus ponderosa* and Associated Species in the South-Western USA," *Journal of Biogeography* 30, no. 2 (2003): 257–276.

58. Wright and Bailey, *Fire Ecology;* Brown, "Biotic Communities of the American Southwest."

59. Wright and Bailey, *Fire Ecology;* Brown, "Biotic Communities of the American Southwest."

60. Brown, "Biotic Communities of the American Southwest."

61. Ibid.

62. R. Matthew Beaty and Alan H. Taylor, "Spatial and Temporal Variation of Fire Regimes in a Mixed Conifer Forest Landscape, Southern Cascades, California, USA," *Journal of Biogeography* 28, no. 8 (2001): 955–966.

63. Brown, "Biotic Communities of the American Southwest"; Beaty and Taylor, "Spatial and Temporal Variation of Fire Regimes."

64. Wright and Bailey, *Fire Ecology.*

65. Harold Weaver, "Effects of Fire on Temperate Forests: Western United States," in *Fire and Ecosystems*, ed. T. T. Koslowski and C. E. Ahlgren, pp. 279–360 (New York: Academic Press, 1974).

66. Christopher H. Baisan and Thomas W. Swetnam, "Fire History on a Desert Mountain Range: Rincon Mountain Wilderness, USA," *Canadian Journal of Forest Research* 20 (1990): 1559–1569; Peter Z. Fulé, Joseph E. Crouse, Thomas A. Heinlein, Margaret M. Moore, W. Wallace Covington, and Greg Verkamp, "Mixed-Severity Fire Regime in a High-Elevation Forest of Grand Canyon, USA," *Landscape Ecology* 18 (2003): 465–486.

67. Beaty and Taylor, "Spatial and Temporal Variation of Fire Regimes."

68. Wright and Bailey, *Fire Ecology.*

69. Jerry F. Franklin and C. T. Dyrness, *The Natural Vegetation of Washington and Oregon*, General Technical Report PNW-8 (Portland, OR: U.S. Department of Agriculture, Forest Service, Pacific Northwest Forest and Range Experiment Station, 1973); Thomas A. Spies and Jerry F. Franklin, "Old Growth and Forest Dynamics in the Douglas-Fir Region of Western Oregon and Washington," *Natural Areas Journal* 8 (1988): 190–201; Agee, *Fire Ecology of Pacific Northwest Forests*; Colin J. Long, Cathy Whitlock, Patrick J. Bartlein, and Sarah H. Millspaugh, "A 9000-Year Fire History from the Oregon Coast Range, Based on a High-Resolution Charcoal Study," *Canadian Journal of Forest Research* 28 (1998): 774–787.

70. P. Alaback, T. T. Veblen, C. Whitlock, A. Lara, T. Kitzberger, and R. Villaba, "Climatic and Human Influences on Fire Regimes in Temperate Forest Ecosystems in North and South America," in *How Landscapes Change: Ecological Studies*, vol. 162, ed. G. A. Bradshaw and P. A. Marquet, pp. 49–87 (Berlin: Springer-Verlag, 2003).

71. Wright and Bailey, *Fire Ecology.*

72. Ibid.

73. Agee, *Fire Ecology of Pacific Northwest Forests.*

74. Ibid. Agee posits an average fire return interval of 230 years but also notes that high spatial variability limits the usefulness of this statistic. Based on Wright and Bailey's discussion, in *Fire Ecology,* of mixed fire regimes in western hemlock–Douglas-fir forests, we might also consider an average fire return interval to be misleading because of high temporal variability.

75. Ibid.

2. FANNING THE FLAMES

1. This variability indicates, at least in part, the important role of weather variables.

2. Data for figures 3 and 4 come from "Fire Information—Wildland Fire Statistics," n.d., National Interagency Fire Center, www.nifc.gov/stats/fires_acres.html, last accessed October 27, 2007.

3. It is important to note that these data reflect trends only at the national level and may obscure local or regional trends. For example, it is possible that a significant increase in fire frequency in one region is simply counterbalanced by a decrease in another region.

4. Data for figure 5 comes from National Interagency Fire Center, "Wildland Fire Accidents by Year," n.d.; it was released by the National Wildfire Coordinating Group's Safety and Health Working Team and is available at www.nifc.gov/reports, last accessed May 31, 2007.

5. S. F. Arno, "Forest Fire History in the Northern Rockies," *Journal of Forestry* 78 (1980): 460–465; J. N. Mast, P. Z. Fulé, M. M. Moore, W. W. Covington, and A. E. M. Waltz, "Restoration of Presettlement Age Structure of an Arizona Ponderosa Pine Forest," *Ecological Applications* 9 (1999): 228–239; M. M. Moore, W. W. Covington, and P. Z. Fulé, "Reference Conditions and Ecological Restoration: A Southwestern Ponderosa Pine Perspective," *Ecological Applications* 9 (1999): 1266–1277; G. Thomas Zimmerman, "Fuels and Fire Behavior," in *Ecological Restoration of Southwestern Ponderosa Pine Forests,* ed. Peter Friederier, pp. 126–143 (Washington, D.C.: Island Press, 2003).

6. The Weeks Act authorized the purchase of large tracts of private land to form a national forest system. John B. Loomis, *Integrated Public Lands Management*, 2nd ed. (New York: Columbia University Press, 2002),

7. National Interagency Fire Center, "Wildland Fire Statistics," n.d., available from www.nifc.gov/fire_info/historical_stats.htm, last accessed October 27, 2007.

8. Denise Gess and William Lutz, *Firestorm at Peshtigo: A Town, Its People, and the Deadliest Fire in American History* (New York: Henry Holt, 2002).

9. In 2004, for example, wildland fires burned more than 6 million acres in Alaska. National Interagency Fire Center, "Highlights for the 2004 Wildland Fire Season," n.d., available from www.nifc.gov/stats/summaries/summary_2004.html, last accessed October 27, 2007.

10. We explore the causes and effects of fire suppression policies in a later chapter.

11. Thomas T. Veblen, "Key Issues in Fire Regime Research for Fuels Management and Ecological Restoration," in *Fire, Fuel Treatments, and Ecological Restoration: Conference Proceedings*, Phillip N. Omi and Linda A. Joyce, technical editors, pp. 259–275, RMRS-P-29 (Fort Collins, CO: U.S. Department of Agriculture, Forest Service, Rocky Mountain Research Station, 2003).

12. Robert A. Heinlein, *Time Enough for Love* (New York: G. P. Putnam's Sons, 1973), p. 371.

13. Intergovernmental Panel on Climate Change, *IPCC Third Assessment Report: Climate Change 2001: Synthesis Report*, ed. R. T. Watson and the Core Writing Team (Cambridge: Cambridge University Press, 2001).

14. Stephen J. Pyne, *Fire in America: A Cultural History of Wildland and Rural Fire* (Seattle: University of Washington Press, 1997), p. 38.

15. H. C. Humphries and P. S. Bourgeron, "Environmental Responses of *Pinus ponderosa* and Associated Species in the South-Western USA," *Journal of Biogeography* 30 (2003): 257–276.

16. Intergovernmental Panel on Climate Change, *IPCC Third Assessment Report: Climate Change 2001: Impacts, Adaptation, and Vulnerability*, ed. James J. McCarthy, Oswaldo F. Canziani, Neil A. Leary, David J. Dokken, and Kasey S. White (Cambridge: Cambridge University Press), 2001.

17. M. D. Flannigan, B. J. Stocks, and B. M. Wotton, "Climate Change and Forest Fires," *Science of the Total Environment* 262 (2000): 221–229; Timothy J. Brown, Beth L. Hall, and Anthony L. Westerling, "The Impact of Twenty-First Century Climate Change on Wildland Fire Danger in the Western United States: An Applications Perspective," *Climate Change* 62, no. 1–3 (2004): 365–388.

18. A. L. Westerling, H. G. Hidalgo, D. R. Cayan, and T. W. Swetnam, "Warming and Earlier Spring Increase U.S. Forest Wildfire Activity," *Science* 313 (2006): 940–943.

19. Jesse A. Logan, Jacques Régniere, and James A. Powell, "Assessing the Impact of Global Warming on Forest Pest Dynamics," *Frontiers in Ecology and the Environment* 1, no. 3 (2003): 130–137.

20. Stanley D. Smith, Travis E. Huxman, Stephen F. Zitzer, Therese N. Charlet, David C. Housman, James S. Coleman, Lynn K. Fenstermaker, Jeffrey R. Seemann, and Robert S. Nowak, "Elevated CO_2 Increases Productivity and Invasive Species Success in an Arid Ecosystem," *Nature* 408 (2000): 79–82.

21. National Assessment Synthesis Team, *Climate Change Impact on the United States: The Potential Consequences of Climate Variability and Change, Report for the US Global Change Research Program* (Cambridge: Cambridge University Press, 2001).

22. Tania Schoennagel, Thomas T. Veblen, and William H. Romme, "The Interaction of Fire, Fuels, and Climate across Rocky Mountain Forests," *BioScience* 34, no. 7 (2004): 661–676.

23. Westerling et al., "Warming and Earlier Spring Increase U.S. Forest Wildfire Activity."

24. Intergovernmental Panel on Climate Change, *Climate Change 2007: The Physical Science Basis. Summary for Policymakers*, ed. Susan Solomon, Dahe Qin, Martin Manning, Melinda Marquis, Kristen

Avenyl, Melinda M. B. Tignor, Henry LeRoy Miller Jr., and Zhenlin Chen, 2007, available from www.ipcc.ch/.

25. Pyne, *Fire in America*.

26. Scott Marshall and Gig Conaughton, "Hunter Charged with Starting Cedar Fire," *North County Times*, October 6, 2004; California Department of Forestry and Fire Protection, "Cedar Fire: Incident Information, Final Update," November 5, 2003, available from www.fire.ca .gov/cdf/incidents/Cedar%20Fire_120/incident_info.html.

27. Wright and Bailey, *Fire Ecology*; Jon E. Keeley and C.J. Fotheringham, "Historic Fire Regime in Southern California Shrublands," *Conservation Biology* 15, no. 6 (2001): 1536–1548, 1542; Jon E. Keeley, C.J. Fotheringham, and M. Morais, "Reexamining Fire Suppression Impacts on Brushland Fire Regimes," *Science* 284 (1999): 1829–1832.

28. Keeley and Fotheringham, "Historic Fire Regime"; Keeley et al., "Reexamining Fire Suppression."

29. Pyne, *Fire in America*.

30. James B. Davis, "The Wildland-Urban Interface: Paradise or Battleground?" *Journal of Forestry* 88, no. 1 (1990): 26–31.

31. Pyne, *Fire in America*.

32. Wildfire Alternatives, "Human Factors of Fire Ignition," 2004, available from http://walter.arizona.edu/society/risk_factors/human_ factors_of_ignition.asp.

33. Davis, "The Wildland-Urban Interface."

34. J.D. Cohen, "A Brief Summary of My Los Alamos Fire Destruction Examination," *Wildfire* 9, no. 4 (2000): 16–18.

35. The five agencies are the Bureau of Land Management, Bureau of Indian Affairs, U.S. Fish and Wildlife Service, National Park Service, and U.S. Department of Agriculture, Forest Service.

36. National Interagency Fire Center, "Suppression Costs for Federal Agencies," n.d., available from www.nifc.gov/stats/suppression_costs .html, last accessed December 7, 2006.

37. U.S. Department of Interior and U.S. Department of Agriculture, *Review and Update of the 1995 Federal Wildland Fire Management Policy, 2001*, available at National Interagency Fire Center, www.nifc.gov/fire

_policy/history/index.htm; National Fire and Aviation Board Policy Directives Task Group, *Federal Wildland Fire Management Policy and Program Review, 1995*, available at www.nifc.gov/fire_policy/95policy.htm.

38. The size of the Ryan Fire and cost of suppression are from University of Arizona, "Huachuca Mountain Fire History, Wildfire Alternatives Project Outline," 2004, available from http://walter.arizona.edu/overview/study_areas/huachuca_fire_hist.asp.

39. John Kupfer, University of Arizona, personal communication, October 3, 2002.

40. "Public Lands: Do We Get What We Pay For?" *TimberLine* (1999), available from www.timberlinemag.com/articledatabase/view.asp?articleID=129.

41. Stephen J. Pyne, Patricia L. Andrews, and Richard D. Laven, *Introduction to Wildland Fire*, 2nd ed. (New York: John Wiley and Sons, 1996), p. 244.

42. Aldo Leopold, "Grass, Brush, Timber, and Fire in Southern Arizona," *Journal of Forestry* 22 (1924): 1–10.

43. James K. Agee and Carl N. Skinner, "Basic Principles of Forest Fuel Reduction Treatments," *Forest Ecology and Management* 211 (2005): 83–96; Harold Weaver, "Effects of Fire on Temperate Forests: Western United States," in *Fire and Ecosystems*, ed. T. T. Kozlowski and C. E. Ahlgren, pp. 279–320 (New York: Academic Press, 1974).

44. Jayne Belnap, "Surface Disturbances: Their Role in Accelerating Desertification," *Environmental Monitoring and Assessment* 37, no. 1–3 (1995): 39–57.

45. E. O. Wooton, "The Range Problem in New Mexico," *New Mexico Agricultural Experiment Station Bulletin* 66 (1908), p. 45.

46. Steve Archer, "Woody Plant Encroachment into Southwestern Grasslands and Savannas: Rates, Patterns, and Proximate Causes," in *Ecological Implications of Livestock Herbivory in the West* (Denver: Society for Range Management, 1994).

47. Robert R. Humphrey, "Fire in the Deserts and Desert Grassland of North America," in *Fire and Ecosystems*, ed. T. T. Koslowski and C. E. Ahlgre, pp. 365–400 (New York: Academic Press, 1974).

48. Steve Archer, "Woody Plant Encroachment into Southwestern Grasslands and Savannas: Rates, Patterns, and Proximate Causes," in *Ecological Implications of Livestock Herbivory in the West*, ed. Martin Vavra, William A. Laycock, and Rexford D. Pieper, pp. 13–68 (Denver, CO: Society for Range Management, 1994); Guy R. McPherson, *Ecology and Management of North American Savannas* (Tucson, AZ: University of Arizona Press, 1997).

49. Wright and Bailey, *Fire Ecology*; McPherson, *Ecology and Management*.

50. S. W. Pacala, G. C. Hurtt, D. Baker, P. Peylin, R. A. Houghton, R. A. Birdsey, L. Heath, E. T. Sundquist, R. F. Stallard, P. Ciais, P. Moorcroft, J. P. Caspersen, E. Shevliakova, B. Moore, G. Kohlmaier, E. Holland, M. Gloor, M. E. Harmon, S. M. Fan, J. L. Sarmiento, C. L. Goodale, D. Schimel, and C. B. Field, "Consistent Land- and Atmosphere-Based U.S. Carbon Sink Estimates," *Science* 292 (2001): 2316–2320.

51. Robert B. Jackson, Jay L. Banner, Esteban G. Jobbagy, William T. Pockman, and Diana H. Wall, "Ecosystem Carbon Loss with Woody Plant Invasion of Grasslands," *Nature* 418 (2002): 623–626.

52. The Environmental Protection Agency provides information on methane. For figures on methane potency, see "Methane," April 27, 2007, www.epa.gov/methane/; for contributions of livestock, see "Ruminant Livestock: Frequent Questions," March 21, 2007, www.epa.gov/methane/rlep/faq.html.

53. In the academic literature, much of this debate can be found in the journal *Conservation Biology*. See, for example, Peter F. Brussard, D. D. Murphy, and C. R. Tracy, "Cattle and Conservation Biology—Another View, *Conservation Biology* 8 (1994): 919–921; Thomas L. Fleischner, "Ecological Costs of Livestock Grazing in Western North America," *Conservation Biology* 8 (1994): 629–644; Reed F. Noss, "Cows and Conservation Biology," *Conservation Biology* 8 (1994): 613–616; George Wuerthner, "Subdivision versus Agriculture," *Conservation Biology* 8 (1994): 905–908; James H. Brown and William McDonald, "Livestock Grazing and Conservation on Southwestern Rangelands," *Conservation*

Biology 9 (1995): 1644–1647; A. Joy Belsky and D. M. Blumenthal, "Effects of Livestock Grazing on Stand Dynamics and Soils in Upland Forests of the Interior West," *Conservation Biology* 11 (1997): 315–327; Carl E. Bock and Jane H. Bock, "Response of Winter Birds to Drought and Short-Duration Grazing in Southeastern Arizona," *Conservation Biology* 13 (1999): 1117–1123; M. L. Floyd, Thomas L. Fleischner, David Hanna, and P. Whitefield, "Effects of Historic Livestock Grazing on Vegetation at Chaco Canyon National Historic Park, New Mexico," *Conservation Biology* 17 (2003): 1703–1711; D. Krueper, J. Bart, and T. D. Rich, "Response of Vegetation and Breeding Birds to the Removal of Cattle on the San Pedro River, Arizona (U.S.A.)," *Conservation Biology* 17 (2003): 607–615; J. D. Maestes, Richard L. Knight, and W. C. Gilgert, "Biodiversity across a Rural Land-Use Gradient," *Conservation Biology* 17 (2003): 1425–1434; Nathan F. Sayre, "The End of the Grazing Debate," *Conservation Biology* 17 (2003): 1186–1188.

54. Jerome E. Freilich, John M. Emlen, Jeffrey J. Duda, D. Carl Freeman, and Philip J. Cafaro, "Ecological Effects of Ranching: A Six-Point Critique," *BioScience* 53 (2003): 759–765.

55. A discussion of the controversy on livestock grazing could fill a book. In fact, it has already filled several. For the interested reader, we recommend the following papers and books: Carl E. Bock, Jane H. Bock, and H. M. Smith, "Proposal for a System of Federal Livestock Exclosures on Public Rangelands in the Western United States," *Conservation Biology* 7 (1993): 731–733; A. Joy Belsky, A., Matzke, and S. Uselman, "Survey of Livestock Influences on Stream and Riparian Ecosystems in the Western United States," *Journal of Soil and Water Conservation* 54 (1999): 419–431; Thomas J. Stohlgren, L. D. Schell, and B. Vanden Heuvel, "How Grazing and Soil Quality Affect Native and Exotic Plant Diversity in Rocky Mountain Grasslands," *Ecological Applications* 9 (199): 45–64; A. Jones, "Effects of Cattle Grazing on North American Arid Ecosystems: A Quantitative Review," *Western North American Naturalist* 60 (2000): 155–164; J. Boone Kauffman and David A. Pyke, "Range Ecology, Global Livestock Influences," in *Encyclopedia of Biodiversity*, vol. 5, *R-Z*, ed. S. A. Levin, pp. 33–52 (San Diego: Academic Press, 2001);

J. D. Maestes, Richard L. Knight, and W. C. Gilgert, "Biodiversity and Land-Use Change in the American Mountain West," *Geographical Review* 91 (2001): 509–524; Jerome E. Freilich, John M. Emlen, Jeffrey J. Duda, D. Carl Freeman, and Philip J. Cafaro, "Ecological Effects of Ranching: A Six-Point Critique," *BioScience* 53 (2003): 759–765; D. Ferguson and N. Ferguson, *Sacred Cows at the Public Trough* (Bend, OR: Maverick Publications, 1983); D. L. Donahue, *The Western Range Revisited* (Norman: University of Oklahoma Press, 1999); Nathan F. Sayre, *The New Ranch Handbook: A Guide to Restoring Western Rangeland* (Santa Fe, NM: Quivira Coalition, 2001); Richard L. Knight, W. C. Gilbert, and Ed Marston, eds., *Ranching West of the 100th Meridian: Culture, Ecology, and Economics* (Washington, DC: Island Press, 2002); George Wuerthner and M. Matteson, eds., *Welfare Ranching: The Subsidized Destruction of the American West* (Washington, DC: Island Press, 2002).

56. Wright and Bailey, *Fire Ecology*, p. 274.

57. Ibid.

58. Ibid.

59. D. C. Donato, J. B. Fontaine, J. L. Campbell, W. D. Robinson, J. B. Kauffman, and B. E. Law, "Post-Wildfire Logging Hinders Regeneration and Increases Fire Risk," *Science* 311 (2006): 352.

60. David R. Foster and David A. Orwig, "Preemptive and Salvage Harvesting of New England Forests: When Doing Nothing Is a Viable Alternative," *Conservation Biology* 20, no. 4 (2006): 959–970.

61. D. B. Lindenmayer and R. F. Noss, "Salvage Logging, Ecosystem Processes, and Biodiversity Conservation," *Conservation Biology* 20, no. 4 (2006): 949–958; J. F. Franklin and J. K. Agee, "Forging a Science-Based National Forest Fire Policy," *Issues in Science and Technology* 20 (2003): 59–66; Richard L. Hutto, "Toward Meaningful Snag-Management Guidelines for Postfire Salvage Logging in North American Conifer Forests," *Conservation Biology* 20, no. 4 (2006): 984–993; Gordon H. Reeves, Peter A. Bisson, Bruce E. Rieman, and Lee E. Benda, "Postfire Logging in Riparian Areas," *Conservation Biology* 20, no. 4 (2006): 994–1004.

62. Carla A. D'Antonio and Peter J. Vitousek, "Biological Invasions by Exotic Grasses, the Grass/Fire Cycle, and Global Change," *Annual Review of Ecology and Systematics* 23 (1992): 63–87.

63. Wright and Bailey, *Fire Ecology.*

64. Jeffrey E. Lovich and David Bainbridge, "Anthropogenic Degradation of the Southern California Desert Ecosystem and Prospects for Natural Recovery and Restoration," *Environmental Management* 24, no. 3 (1999): 309–326; Wright and Bailey, *Fire Ecology;* Humphrey, "Fire in the Deserts and Desert Grassland of North America."

65. David E. Brown and Richard A. Minnich, "Fire and Changes in Creosote Bush Scrub of the Western Sonoran Desert, California," *American Midland Naturalist* 116, no. 2 (1986): 411–422.

66. D'Antonio and Vitousek, "Biological Invasions."

67. Brown and Minnich, "Fire and Changes in Creosote Bush Scrub"; Steven P. McLaughlin and Janice E. Bowers, "Effects of Wildfire on a Sonoran Desert Plant Community," *Ecology* 63, no. 1 (1982): 246–248.

68. D'Antonio and Vitousek, "Biological Invasions."

69. Brown and Minnich, "Fire and Changes in Creosote Bush Scrub."

70. D'Antonio and Vitousek, "Biological Invasions"

71. Humphrey, "Fire in the Deserts and Desert Grassland of North America."

72. Brown and Minnich, "Fire and Changes in Creosote Bush Scrub."

73. Veblen, "Key Issues in Fire Regime Research."

3. THE FAILED STATE OF FIRE SUPPRESSION

1. Bradley C. Karkkainen, "Collaborative Ecosystem Governance: Scale, Complexity, and Dynamism," *Virginia Environmental Law Journal* 21 (2001–2002): 189–243.

2. National Interagency Fire Center, "Historically Significant Wildland Fires," available from www.nifc.gov/fire_info/historical.stats .htm.

3. Mitch Tobin, "Rodeo-Chediski Fire Anniversary: Signs of Hope Sprout in Wake of Fire," *Arizona Daily Star,* June 6, 2005.

4. U.S. Department of Agriculture, Forest Service, "Apache-Sitgreaves and Tonto National Forests Rodeo/Chediski Fire Salvage and Rehabilitation Project," *Federal Register* 67, no. 187 (2002): 60637–60639, 60638.

5. William L. Baker and Donna Ehle, "Uncertainty in Surface-Fire History: The Case of Ponderosa Pine Forests in the Western United States," *Canadian Journal of Forest Research* 33 (2003): 1205–1226, 1223; William L. Baker and Donna Ehle, "Uncertainty in Fire History and Restoration of Ponderosa Pine Forests in the Western United States," in *Fire, Fuel Treatments, and Ecological Restoration: Conference Proceedings,* Philip N. Omi and Linda A. Joyce, technical editors, pp. 319–333, Proceedings RMRS-P-29 (Fort Collins, CO: U.S. Department of Agriculture, Forest Service, Rocky Mountain Research Station, 2003).

6. See, among many others, W. Wallace Covington, Peter Z. Fulé, Margaret M. Moore, Stephen C. Hart, Thomas E. Kolb, Joy N. Mast, Stephen S. Sackett, and Richard R. Wagner, "Restoring Ecosystem Health in Ponderosa Pine Forests of the Southwest," *Journal of Forestry* 95, no. 4 (1997): 23–29; W. Wallace Covington, "The Evolutionary and Historical Context," in *Ecological Restoration of Southwestern Ponderosa Pine Forests,* ed. Peter Friederici, pp. 26–47 (Washington, DC: Island Press, 2003); Peter Z. Fulé, W. Wallace Covington, and Margaret M. Moore, "Determining Reference Conditions for Ecosystem Management of Southwestern Ponderosa Pine Forests," *Ecological Applications* 7 (1997): 895–908; Peter Z. Fulé, Charles McHugh, Thomas A. Heinlein, and W. Wallace Covington, "Potential Fire Behavior Is Reduced Following Forest Restoration Treatments," in *Ponderosa Pine Ecosystems Restoration and Conservation: Steps toward Stewardship,* comp. Regina K. Vance, C. B. Edminster, W. W. Covington, and J. A. Blake, pp. 28–35, Proceedings RMRS-P-22 (Fort Collins, CO: U.S. Department of Agriculture, Forest Service, Rocky Mountain Research Station, 2001); Peter Z. Fulé, Amy E. M. Waltz, W. Wallace Covington, and Thomas A. Heinlein, "Measuring Forest Restoration Effectiveness

in Reducing Hazardous Fuels," *Journal of Forestry* 99, no. 11 (2001): 24–29.

7. Mark A. Finney, Charles W. McHugh, and Isaac C. Grenfell, "Stand- and Landscape-Level Effects of Prescribed Burning on Two Arizona Wildfires," *Canadian Journal of Forest Research* 35 (2005): 1714–1722.

8. Stephen Pyne, a fire historian at Arizona State University, has written such comprehensive and insightful descriptions of the history of fire and fire management in the United States that our attempts here are somewhat redundant. Nonetheless, we feel a brief review of this chapter of history is crucial to understanding our national failure in managing fire. This historical discussion draws heavily on pages 242–250 of Pyne's *Fire in America: A Cultural History of Wildland and Rural Fire*, and we are heavily indebted to his exhaustive research (Seattle: University of Washington Press, 1997).

9. Randal O'Toole, "Money to Burn: Wildfire and the Budget," in *The Wildfire Reader: A Century of Failed Forest Policy*, ed. George Wuerthner, pp. 250–261 (Washington, DC: Island Press, 2006).

10. Henri D. Grissino-Mayer and Thomas W. Swetnam, "Century-Scale Climate Forcing of Fire Regimes in the American Southwest," *The Holocene* 10, no. 2 (2000): 213–220; Schoennagel et al., "Interaction of Fire, Fuels, and Climate.

11. Backer and others provide an overview of the environmental impacts of fire suppression activities: D. M. Backer, S. E. Jensen, and G. R. McPherson, "Impacts of Fire-Suppression Activities on Natural Communities," *Conservation Biology* 18, no. 4 (2004): 937–946.

12. K. J. Buhl and S. J. Hamilton, "Acute Toxicity of Fire-Retardant and Foam Suppressant Chemicals to Early Life Stages of Chinook Salmon *(Omcorhynchus tshawytscha)*," *Environmental Toxicology and Chemistry* 17 (1998): 1589–1599.

13. Some fire-retardant chemicals that are fairly inert in their original form can actually release toxic cyanide when exposed to a certain level of ultraviolet radiation. E. E. Little and R. D. Calfee, "Environmental Implications of Fire-Retardant Chemicals: Project Summary," 2002,

U.S. Department of the Interior, U.S. Geological Survey, available from www.cerc.cr.usgs.gov/pubs/center/pdfdocs/eco-03.pdf.

14. M. P. Gaikowski, S. J. Hamilton, K. J. Buhl, S. F. McDonald, and C. H. Summers, "Acute Toxicity of Three Fire-Retardant and Two Fire-Suppressant Foam Formulations to the Early Life Stages of Rainbow Trout *(Oncorhynchus mykiss),*" *Environmental Toxicology and Chemistry* 15 (1996): 1365–1374.

15. Independent Review Board, "Cerro Grande Prescribed Fire: Independent Review Board Report," May 26, 2000, available from www.nps.gov/cerrogrande/; National Park Service, "Cerro Grande Prescribed Fire Investigation Report: Delivered to the Secretary of Interior, Bruce Babbitt on May 18, 2000," 2000, available from: www.nps.gov/cerrogrande/.

16. General Accounting Office, *Western National Forests: A Cohesive Strategy Is Needed to Address Catastrophic Wildfire Threats* GAO/RCED-99–65 (Washington, DC, 1999).

17. Timothy Ingalsbee, "Money to Burn: The Economics of Fire Suppression," n.d., available from www.fire-ecology.org/citizen/money_to_burn.htm, last accessed October 27, 2007.

18. Office of Management and Budget, *Budget of the United States Government, Fiscal Year 2003,* 2003, p. 66, available from www.whitehouse.gov/omb/budget/fy2003/budget.html.

19. This policy was later extended to include some "presuppression" activities, including equipment purchases. O'Toole, "Money to Burn."

20. Ibid.

21. Indeed, the National Fire Plan seriously overstates the suppression–fuel buildup link, because it overgeneralizes it to all North American forests.

4. LOGGING THE FORESTS TO SAVE THEM

1. Z. Gedalof, D. L. Peterson, and N. J. Mantua, "Atmospheric, Climatic, and Ecological Controls on Extreme Wildfire Years in the Northwestern United States," *Ecological Applications* 15, no. 1 (2005): 154–174.

2. Russell T. Graham, technical editor, *Hayman Fire Case Study: Summary*, General Technical Report RMRS-GTR-115 (Ogden, UT: U.S. Department of Agriculture, Forest Service, Rocky Mountain Research Station, 2003).

3. Ibid.

4. Tania Schoennagel, Thomas T. Veblen, and William H. Romme, "The Interaction of Fire, Fuels, and Climate across Rocky Mountain Forests. *BioScience* 34, no. 7 (2004): 661–676.

5. Office of the President, "Healthy Forests: An Initiative for Wildfire Prevention and Stronger Communities," press release of August 22, 2002, available at www.whitehouse.gov/infocus/healthyforests/Healthy _Forests_v2.pdf.

6. Western ecosystems that likely experienced large, intense, "catastrophic" fires before widespread fire suppression include mixed-conifer and spruce-fir forests, some low-elevation forests in the Pacific Northwest and Alaska, chaparral, lodgepole pine forests, and even some ponderosa pine forests. See, for example, James K. Agee, *Fire Ecology of Pacific Northwest Forests* (Washington, DC: Island Press, 1993); Edward A. Johnson, *Fire and Vegetation Dynamics: Studies from the North American Boreal Forest* (New York: Cambridge University Press, 1992); Edward A. Johnson and Kiyoko Miyanishi, eds., *Forest Fires: Behavior and Ecological Effects* (New York: Academic Press, 2001); Theodore Thomas Kozlowski and Clifford Elmer Ahlgren, *Fire and Ecosystems* (New York: Academic Press, 1974); Linda L. Wallace, *After the Fires: The Ecology of Change in Yellowstone National Park* (New Haven, CT: Yale University Press, 2004); Robert J. Whelan, *The Ecology of Fire* (New York: Cambridge University Press, 1995).

7. This is the Fire Regime Condition Class Website (www.frcc.gov), an interagency tool first developed in 2003 by Wendel Hann, Doug Havlina, Ayn Shlisky, et al. and sponsored by the Nature Conservancy; U.S. Department of Agriculture, Forest Service; U.S. Department of the Interior; the Nature Conservancy; and Systems for Environmental Management.

8. C. C. Hardy, K. M. Schmidt, J. M. Menakis, and N. R. Samson, "Spatial Data for National Fire Planning and Fuel Management," *International Journal of Wildland Fire* 10 (2001): 353–372; K. M. Schmidt,

J. P. Menakis, C. C. Hardy, W. J. Hann, and D. L. Bunnell, *Development of Coarse-Scale Spatial Data for Wildland Fire and Fuel Management*, General Technical Report RMRS-GTR-87 (Fort Collins, CO: U.S. Department of Agriculture, Forest Service, Rocky Mountain Research Station, 2002).

9. U.S. Department of Agriculture, Forest Service, "The Process Predicament: How Statutory, Regulatory, and Administrative Factors Affect National Forest Management," June 2002, available from www .fs.fed.us/projects/documents/Process-predicament.pdf.

10. U.S. Department of the Interior and U.S. Department of Commerce, "Joint Counterpart Endangered Species Act Section 7 Consultation Regulations (Final Rule)," *Federal Register* 68, no. 235 (2003): 68254–68285; U.S. Department of Agriculture and U.S. Department of the Interior, "National Environmental Policy Act Documentation Needed for Fire Management Activities: Categorical Exclusions (Final Rule)," *Federal Register* 68, no. 108 (2003): 33814–33824.

11. F. C. Ford-Robertson, *Terminology of Forest Science, Technology Practice, and Products* (Washington, DC: Society of American Foresters, 1971); Ralph D. Nyland, *Silviculture: Concepts and Applications* (New York: McGraw-Hill, 1996); David M. Smith, Bruce C. Larson, Matthew J. Kelty, and Mark S. Ashton, *The Practice of Silviculture: Applied Forest Ecology* (New York: Wiley, 1997).

12. See, for example, P. Z. Fulé, W. W. Covington, H. B. Smith, J. D. Springer, T. A. Heinlein, K. D. Huisinga, and M. M. Moore, "Comparing Ecological Restoration Alternatives: Grand Canyon, Arizona," *Forest Ecology and Management* 170 (2002): 19–41; J. N. Mast, P. Z. Fulé, M. M. Moore, W. W. Covington, and A. E. M. Waltz, "Restoration of Presettlement Age Structure of an Arizona Ponderosa Pine Forest," *Ecological Applications* 9, no. 1 (1999): 228–239.

13. For example, U.S. Department of Agriculture and U.S. Department of the Interior, "National Environmental Policy Act Documentation Needed for Fire Management Activities: Categorical Exclusions (Final Rule)," *Federal Register* 68, no. 108 (2003): 33814–33824, 33814.

14. Healthy Forests Restoration Act of 2003, Pub. L. No. 108–148, 117 Stat. 1887, Sec. 102(f)1, emphasis added.

15. Erik J. Martinson and Philip N. Omi, "Performance of Fuel Treatments Subjected to Wildfires," in *Fire, Fuel Treatments, and Ecological Restoration*, P. N. Omi and L. A. Joyce, technical editors, pp. 7–13, Proceedings RMRS-P-29 (Fort Collins, CO: U.S. Department of Agriculture, Forest Service, 2003).

16. See, for example, Merrill R. Kaufmann, Laurie S. Huckaby, Paula Jane Fornwalt, Jason M. Stoker, and William H. Romme, "Using Tree Recruitment Patterns and Fire History to Guide Restoration of an Unlogged Ponderosa Pine/Douglas-Fir Landscape in the Southern Rocky Mountains after a Century of Fire Suppression," *Forestry* 76, no. 2 (2003): 231–241; J. D. McIver, P. W. Adams, J. A. Doyal, E. S. Dres, B. R. Hartsough, L. D. Kellogg, C. G. Niwa, R. Ottmar, R. Peck, M. Taratoot, T. Togerson, and A. Youngblood, "Environmental Effects and Economics of Mechanized Logging for Fuel Reduction in Northeastern Oregon Mixed-Conifer Stands," *Western Journal of Applied Forestry* 18, no. 4 (2003): 238–249; R. M. Muzika, S. T. Gruschecky, A. M. Liebhold, and R. L. Smith, "Using Thinning as a Management Tool for Gypsy Moth: The Influence on Small Mammal Abundance," *Forest Ecology and Management* 192, no. 203 (2003): 349–359; Ralph D. Nyland, *Silviculture: Concepts and Applications* (New York: McGraw Hill, 1996).

17. David A. Perry, *Forest Ecosystems* (Baltimore: Johns Hopkins University Press, 1994), p. 490.

18. B. Pinel-Alloul, E. Prepas, D. Planas, R. Steedman, and T. Charette, "Watershed Impacts of Logging and Wildfire: Case Studies in Canada," *Lake and Reservoir Management* 18 (2002): 307–318, 317.

19. Chris Maser, Robert F. Tarrant, James M. Trappe, and Jerry F. Franklin, technical editors, *From the Forest to the Sea: A Story of Fallen Trees*, General Technical Report PNW-GTR-229 (Portland, OR: U.S. Department of Agriculture, Forest Service, Pacific Northwest Research Station, 1988).

20. U.S. Fish and Wildlife Service, "Kirtland's Warbler Wildlife Management Area," n.d., available from http://refuges.fws.gov/profiles/index.cfm?id=31513, last accessed October 27, 2007.

21. U.S. Department of the Interior and U.S. Department of Commerce, "Joint Counterpart Endangered Species Act Section 7 Consultation Regulations (Final Rule)," *Federal Register* 68, no. 108 (June 5, 2003): 33, 806; U.S. Department of Agriculture, "Cabinet Officials Report Progress on President Bush's Healthy Forests Initiative," May 2003, available from www.usda.gov/news/releases/2003/05/0177.htm; Gail A. Norton and Ann M. Veneman, "Cut the Red Tape, Restore Our Forests," op-ed, *Seattle Times*, 22 August 2003.

22. U.S. Department of Agriculture, Forest Service, "The Process Predicament," p. 5.

23. U.S. Department of Agriculture, "Remarks by President George W. Bush at Signing of H.R. 1904, the Healthy Forests Restoration Act of 2003," December 2003, available from www.usda.gov/news/releases/2003/12/hfiremarks.htm.

24. "Fire Prevention Blame Game Is Complex," *Arizona Daily Star,* June 25, 2002.

25. "Forest Service Acts to Improve Firefighter Safety," *The Forestry Source* (Society of American Foresters), November 2001.

26. Bradley C. Karkkainen, "Whither NEPA? *NYU Environmental Law Journal* 12 (2004): 333.

27. For the National Environmental Policy Act, for example, see 40 CFR 1501.8.

28. General Accounting Office, *Forest Service: Information on Decisions Involving Fuels Reduction Activities,* GAO-03–0689R (Washington, DC, 2003), p. 13.

29. General Accounting Office, *Wildland Fire Management: Additional Actions Required to Better Identify and Prioritize Lands Needing Fuels Reduction,* GAO-03–805 (Washington, DC, 2003), p. 9.

30. Hanna J. Cortner and Margaret A. Moote, *The Politics of Ecosystem Management* (Washington, DC: Island Press, 1999).

31. 36 CFR 215.

32. White House Office of Communications, "Healthy Forests: An Initiative for Wildfire Prevention and Stronger Communities," available from www.whitehouse.gov/infocus/healthyforests/restor-act-pg2.html.

33. Jerry F. Franklin and James K. Agee, "Forging a Science-Based National Forest Fire Policy," *Issues in Science and Technology* 20, no. 1 (2003): 59–66, quote on p. 60.

34. Office of the President, "Executive Order 12866 of September 30, 1993: Regulatory Planning and Review," *Federal Register* 58, no. 190 (1993): 51735–51744, 51735.

35. 36 CFR 215. Several citizen groups, led by the environmental nongovernmental organization EarthJustice, are currently challenging these changes at the U.S. District Court level, in Montana.

36. Mike Anderson, from a memo to the public on Forest Service appeal regulations, July 4, 2003, available from www.wilderness.org/Library/documents/upload/Analysis-of-new-Appeals-Reform-Act-regulations.pdf.

37. 36 CFR 215.4, 215.12.

38. *Forest Service Handbook* 1909.15, Chapter 30, section 31.

39. 36 CFR 215.2

40. Committee on the Judiciary, House of Representatives, "Report Together with Dissenting Views to Accompany H.R. 1904," 108th Congress, 1st sess., Rept. 108–96 (Washington, DC, May 16, 2003).

41. Ibid., p. 84.

42. Sharon Buccino, "NEPA Under Assault: Congressional and Administrative Proposals Would Weaken Environmental Review and Public Participation," *NYU Environmental Law Journal* 12 (2003): 50–73, 50.

43. Ibid., p. 10.

44. Anne Larason Schneider and Helen Ingram, *Policy Design for Democracy* (Lawrence: University Press of Kansas, 1997), p. 5.

45. U.S. Department of Agriculture, Forest Service, "The Process Predicament," p. 28.

46. Office of the Press Secretary, "President Announces Healthy Forests Initiative: Remarks by the President on Forest Health and Preservation, Central Point, Oregon" (Washington, DC, August 22, 2002), p. 1.

47. Jacqueline Vaughn and Hanna J. Cortner, *George W. Bush's Healthy Forests: Reframing the Environmental Debate* (Boulder: University Press of Colorado, 2005), p. 125.

48. Healthy Forests Restoration Act of 2003, p. 15.

5. TOOLS FOR LIVING WITH FIRE

1. Loomis, *Integrated Public Lands Management: Principles and Applications to National Forests, Parks, Wildlife Refuges, and BLM Lands*, 2nd ed. (New York: Columbia University Press, 2002), p. 24.

2. On the very day we write this, three firefighters have been killed in a hopeless attempt to protect an already evacuated home from a fast-moving fire in California chaparral east of Los Angeles. We feel strongly that firefighters should not be asked to risk their lives to protect property when other human lives are not at risk.

3. General Accounting Office, *Recreation Fees*, RCED-99-7 (Washington, DC: GAO, 1998), p. 15.

4. As we noted earlier, from an ecological perspective there is relatively little difference between a lightning-ignited fire and a human-ignited fire. At this time, however, federal management agencies will not allow arson fires to burn.

5. Stephen J. Pyne, Patricia L. Andrews, and Richard D. Laven, *Introduction to Wildland Fire* (New York: Wiley, 1996).

6. Theodore W. Daniel, John A. Helms, and Frederick S. Baker, *Principles of Silviculture*, 2nd ed. (New York: McGraw-Hill Book Company, 1979).

7. Russell T. Graham, Sarah McCaffrey, and Theresa B. Jain, technical editors, *Science Basis for Changing Forest Structure to Modify Wildfire Behavior and Severity*, General Technical Report RMRS-GTR-120 (Fort Collins, CO: U.S. Department of Agriculture, Forest Service, Rocky Mountain Research Station, 2004).

8. Thomas T. Veblen, "Key Issues in Fire Regime Research for Fuels Management and Ecological Restoration," in *Fire, Fuel Treatments, and Ecological Restoration: Conference Proceedings*, Philip N. Omi and Linda A. Joyce, technical editors, pp. 259–275, Proceedings RMRS-P-29 (Fort Collins, CO: U.S. Department of Agriculture, Forest Service, Rocky Mountain Research Station, 2003).

9. Dennis L. Lynch and Kurt Mackes, "Costs for Reducing Fuels in Colorado Forest Restoration Projects," in *Fire, Fuel Treatments, and*

Ecological Restoration: Conference Proceedings, Philip N. Omi and Linda A. Joyce, technical editors, pp. 167–175, Proceedings RMRS-P-29 (Fort Collins, CO: U.S. Department of Agriculture, Forest Service, Rocky Mountain Research Station, 2003), p. 168.

10. Kirsten M. Schmidt, James P. Menakis, Colin Hardy, Wendel S. Hann, and David L. Bunnell, *Development of Coarse-Scale Spatial Data for Wildland Fire and Fuel Management*, RMRS-0-TR87 (Fort Collins, CO: U.S. Department of Agriculture, Forest Service, Rocky Mountain Research Station, 2002), p. 14.

11. Graham et al., *Science Basis for Changing Forest Structure*.

12. Jack D. Cohen, "Preventing Disaster: Home Ignitability in the Wildland-Urban Interface," *Journal of Forestry* 98, no. 3 (2000): 15–21; James B. Davis, "The Wildland-Urban Interface: Paradise or Battleground?" *Journal of Forestry* 88, no. 1 (1990): 26–31.

13. Stephen J. Pyne, *Fire in America: A Cultural History of Wildland and Rural Fire* (Seattle: University of Washington Press, 1997), p. 17.

14. Hanna J. Cortner, Malcolm J. Zwolinski, Edwin H. Carpenter, and Jonathan G. Taylor, "Public Support for Fire-Management Policies," *Journal of Forestry* 82, no. 6 (1984): 359–361.

15. Philip D. Gardener, Hanna J. Cortner, Keith F. Widaman, and Kathryn J. Stenberg, "Forester Attitudes toward Alternative Fire Management Policies," *Environmental Management* 9, no. 4 (1985): 303–312.

16. Katie Kneeshaw, Jerry J. Vaske, Alan D. Bright, and James D. Absher, "Situational Influences of Acceptable Wildland Fire Management Actions," *Society and Natural Resources* 17, no. 6 (2004): 477–489.

17. Allen A. Steuter and Guy R. McPherson, "Fire as a Physical Stress," in *Wildland Plants: Physiological Ecology and Developmental Morphology*, ed. Donald J. Bedunah and Ronald E. Sosabee, pp. 550–579 (Denver, CO: Society for Range Management, 1995); Harold Weaver, "Effects of Fire on Temperate Forests: Western United States," in *Fire and Ecosystems*, ed. T. T. Kozlowski and C. E. Ahlgren, pp. 279–319 (New York: Academic Press, 1974).

18. Ralph D. Nyland, *Silviculture: Concepts and Applications* (New York: McGraw Hill, 1996); Weaver, "Effects of Fire on Temperate Forests."

19. D. A. Perala, "*Populus temuloides* Michx. Quaking Aspen," in *Silvics of North America*, vol. 2, *Hardwoods*, Russell M. Burns and Barbara H. Honkala, technical coordinators, pp. 555–569, Agriculture Handbook 654 (Washington, DC: U.S. Department of Agriculture, Forest Service, 1990).

20. Paulo M. Fernandez and Herminio Botelho, "A Review of Prescribed Burning Effectiveness in Fire Hazard Reduction," *International Journal of Wildland Fire* 12 (2003): 117–128.

21. Pyne, *Fire in America*, pp. 330–331.

22. See "The Effect of Wildland Fire on Aquatic Ecosystems in the Western USA," special issue, *Forest Ecology and Management* 178 (2003): 1–2.

23. Richard Feynman, "Final Report on the *Challenger* Space Shuttle Explosion," Appendix F, "Report of the Presidential Commission on the Space Shuttle Challenger Accident," 1986, available from http://science.ksc.nasa.gov/shuttle/missions/51-1/docs/rogers-commission/Appendix-F.txt.

6. POLICY SOLUTIONS

1. Jerry F. Franklin and James K. Agee, "Forging a Science-Based National Forest Fire Policy," *Issues in Science and Technology* 20, no. 1 (2003): 59–66, 59.

2. The number of acres burned was, at the time, the highest on the National Interagency Fire Center's records, which begin in 1960. In 2005, 8,686,753 acres burned, and the record was broken. National Interagency Fire Center, "Fire Information—Wildland Fire Statistics," n.d., available from www.nifc.gov/stats/fires_acres.html, last accessed October 27, 2007.

3. National Interagency Fire Center, "Historically Significant Wildland Fires," n.d., available from www.nifc.gov/stats/historicalstats.html, last accessed October 27, 2007.

4. U.S. Department of the Interior and U.S. Department of Agriculture, *Federal Wildland Fire Management Policy of 1995* (Washington, DC, 1995).

5. National Interagency Fire Center, "Wildland Fire Accidents by Year," n.d., available from www.nifc.gov/reports/, last accessed May 31, 2007.

6. U.S. Department of the Interior and U.S. Department of Agriculture, *Federal Wildland Fire Management Policy of 1995*. This section closely follows Sara E. Jensen, "Policy Tools for Wildland Fire Management: Principles, Incentives, and Conflicts," *Natural Resources Journal* 46, no. 4 (2006): 959–1003.

7. U.S. Department of the Interior and U.S. Department of Agriculture, *Federal Wildland Fire Management Policy of 1995*, n.p. (p. 8).

8. U.S. Department of the Interior and U.S. Department of Agriculture, *Review and Update of the 1995 Federal Wildland Fire Management Policy*, 2001, available from www.nifc.gov/fire_policy/index.htm.

9. Government Accountability Office, *Wildland Fire Management: Update on Federal Agency Efforts to Develop a Cohesive Strategy to Address Wildland Fire Threats*, GAO-06-671R (Washington, DC, 2006).

10. Many fire management plans can be accessed online. For example, the plan for Yosemite National Park is available at www.nps.gov/archive/yose/planning/fire/.

11. U.S. Department of the Interior and U.S. Department of Agriculture, *Review and Update of the 1995 Federal Wildland Fire Management Policy*.

12. U.S. Department of the Interior and U.S. Department of Agriculture, "A Report to the President in Response to the Wildfires of 2000: Managing the Impact of Wildfire on Communities and the Environment," 2000, available from www.fireplan.gov/reports/8-20-en.pdf.

13. Wilderness Society, *The Federal Wildland Fire Budget*, 2004, p. 2, available from www.wilderness.org/Library/Documents/upload/The-Federal-Wildland-Fire-Budget-Let-s-Prepare-Not-Just-React.pdf; U.S. Department of the Interior and U.S. Department of Agriculture, *Federal Wildland Fire Management Policy of 1995*.

14. Government Accountability Office, *Wildland Fire Management*.

15. Steven Rathgeb Smith and Helen Ingram, "Public Policy and Democracy," in *Public Policy for Democracy*, ed. Helen Ingram and Steven

Rathgeb Smith, pp. 1–14 (Washington, DC: Brookings Institution, 1993).

16. Ibid., p. 1.

17. Daniel A. Mazmanian and Paul A. Sabatier, *Implementation and Public Policy* (Glenview, IL: Scott, Foresman, 1983).

18. Smith and Ingram, "Public Policy and Democracy," p. 10.

19. Marcus E. Etheridge, "Procedures for Citizen Involvement in Environmental Policy: An Assessment of Policy Effects," in *Citizen Participation in Public Decision-Making,* ed. Jack DeSario and Stuart Langton, pp. 115–131 (New York: Greenwood Press, 1987).

20. Marc Landy, "Public Policy and Citizenship," in *Public Policy for Democracy,* ed. Helen Ingram and Steven Rathgeb Smith, pp. 19–44 (Washington, DC: Brookings Institution, 1993).

21. Ibid., p. 40.

22. Tomas M. Koontz, *Federalism in the Forest: National versus State Natural Resource Policy* (Washington, DC: Georgetown University Press, 2002).

INDEX

Text: 10/15 Janson
Display: Janson
Compositor: Binghamaton Valley Composition, LLC
Indexer: Thérèse Shere
Illustrator: Bill Nelson
Printed and binder: Maple-Vail Manufacturing Group